ERGONOMICS
MAKING PRODUCTS AND PLACES FIT PEOPLE

ERGONOMICS
MAKING PRODUCTS AND PLACES FIT PEOPLE

KATHLYN GAY

ENSLOW PUBLISHERS, INC.

Bloy St. & Ramsey Ave. P.O. Box 38
Box 777   Aldershot
Hillside, N.J. 07205   Hants GU12 6BP
U.S.A.   U.K.

Copyright ©1986 by Kathlyn Gay

All rights reserved.

No part of this book may be reproduced by any means without the written permission of the publisher.

Library of Congress Cataloging in Publication Data

Gay, Kathlyn.
　　Ergonomics: making products and places fit people.

　　Bibliography: p.
　　Includes index.
　　Summary: Describes the research, tools, and uses of ergonomics or human factor engineering, in which such diverse products as office furniture, cameras, and space suits are designed to suit people's needs.
　　1. Human engineering. [1. Human engineering]
I. Title.
TA166.G39　1986　　　　　620.8'2　　　　85-20634
ISBN 0-89490-118-4

Illustration Credits
Brian Byrn, pp. 18, 19, 21, 22, 23; Nick Simonelli, 3M Company, pp. 29, 113; Steelcase Inc., pp. 35, 95; IBM Corporation, pp. 38, 39; Kathlyn Gay, 41, 42, 45, 50, 53, 60, 61, 63, 68, 94, 100; *Bell System Technical Journal* Copyright 1983, AT&T, pp. 55, 56; Indiana Driver's Manual, Bureau of Motor Vehicles, p. 67; Mercedes-Benz of North America, p. 71; Chevrolet Motor Division, General Motor Corporation, p. 73; NASA, pp. 75, 79, 80, 81, 82, 84, 85, 87, 115, 116; *The Elkhart Truth,* p. 98; VISUALTEK, p. 103; Tom Cannon, p. 104; James Mueller, George Washington University, p. 106; Kenneth Kennedy, Consumer Products Tech Group of the Human Factors Society p. 110; Nautilus Environmedical Systems, p. 117.

ACKNOWLEDGMENTS

A special thanks to the following for their technical advice and/or patient explanations of ergonomics/human factors principles as applied in their areas of expertise: Alphonse Chapanis, Communications Research Laboratory; Harry L. Davis and staff, Human Factors Section, Eastman Kodak Company; Jack A. Laveson, Chairman Consumer Products Technical Group, Human Factors Society; Marian Knowles and staff, Human Factors Society; Steven M. Casey, Principal Scientist, Anacapa Sciences, Inc.; Thomas Bauckham, Senior Industrial Hygienist, Miles Laboratory, Inc.; Kenneth Collister and Steve Johnson, Designers, Ames Division, Miles Laboratory, Inc.; Nick Simonelli, Human Factors Specialist and colleagues at 3M Company; Tom Cannon, Product Design and Engineering Consultant; Susanne Gatchell, Engineer in Charge, and her assistant Virginia Case, Fisher Body Division, GM; James Mueller, Mueller & Zullo, Inc. Consultants; Gerald Botzum, Eli Lilly and Co.; Tom Slager, Slager Systems Division, O/I, Inc.; Rosalie Dolmatch, Coordinator "The Product of Design" Exhibition, The Katonah Gallery; and personnel at Honeywell, AT&T Bell Laboratories, Steelcase, Inc., IBM, Domore Corporation, Cooper-Hewitt Museum (Henry Dreyfuss Archives), George Washington University Rehabilitation Research and Training Center, and the Association for the Disabled of Elkhart County, Inc.

–K.G.

CONTENTS

1. "Ergo-What?" 11
2. Ergonomics in Industry 17
3. "Humanizing" the Office 31
4. Creating Clear Visual Messages 41
5. Designing for Consumers 51
6. Human Factors in Transportation 65
7. Designs From Space Research 77
8. People Space on Earth 91
9. Handiable Designs 99
10. Tools Ergonomists Use 109
    Glossary, Abbreviations, and Addresses 119
    Further Reading 121
    Index 125

FOREWORD

In recent years, designers have become more aware of the needs, abilities, and limitations of consumers, including those with handicaps. They have made significant strides in applying ergonomic principles—consideration of human factors or characteristics—in the design of products, living areas, and workplaces. From pliers with angled handles to user-friendly computer systems, many different products have been designed to make life easier and safer and thus more productive.

A European friend of mine once told me that the reason the French have good cuisine is that the French people would not accept anything less. In the same fashion, knowledgeable consumers will not accept anything less than good ergonomic design wherever people live, play, work, and use manufactured products. This book makes a significant contribution toward developing just such an awareness on the part of the consumer.

Harry L. Davis

Director
Human Factors Section
Eastman Kodak Company

"ERGO-WHAT?" 1

- What can be done to make computers easier for people to use?

- What types of signs and symbols along highways are easiest to read and to understand when visibility is poor?

- What is the best way to train astronauts for lengthy space missions?

- How should robots be designed for use in auto manufacturing plants?

- How much information can an air traffic controller handle and still safely guide pilots and their aircraft?

- What kinds of chairs fit the majority of people who use them in offices or classrooms?

- What is the proper height for a kitchen counter?

These questions pose very different problems, but all have something in common. They represent just a few of the design tasks that are tackled by specialists in the field of *ergonomics.*

"Ergo-what?" you might ask, and with good reason. Although ergonomics is a "buzzword" in some industries, it is not as widely used in the United States and Canada as it is in other parts of the world. In fact, North Americans often call ergonomics work *human factors engineering* or simply *human factors.*

In most cases, ergonomics and human factors can be used synonymously. Both terms mean the science of making machines, products, and places fit people. The terms also refer to a design philosophy that emphasizes the importance of good design so that things are easy, safe, comfortable, and efficient to use.

Professionals in the field, who are known as *ergonomists* or *human factors specialists*, have backgrounds in industrial engineering, psychology, data processing, physiology, medicine, architecture, sociology, and/or other traditional professions. But they specialize in the research and design of products and places to make them suit human needs.

To design things to fit people, ergonomists must pay attention to human factors—human characteristics. They use information from many disciplines and must know, for example, people's sizes and shapes, their abilities to see and hear, their physical strengths and limits, how they handle information, and many other factors.

Knowledge about human factors is essential to make better use of technology for working and living, states the Human Factors Society, the national organization of professionals in the field with headquarters in Santa Monica, California. There are an estimated six thousand human factors professionals in the United States. Worldwide there are about ten thousand. All share the belief that "people should get top priority in the way things are designed."

Is this a new concept? Not exactly. Human factors or ergonomic principles have ancient roots. Over millions of years, people have improved on the designs of simple hand tools and crude shelters so that they better serve human needs. But it was not until the industrial revolution in the late 1700s and through the 1800s that machines and electrical power began to transform the way people worked and lived.

During the last century, scientists studied the way machines helped people do heavy muscular jobs such as making steel, and how machines performed tasks that had to be repeated again and again, as in weaving cloth. At first the studies were concerned with how people could adapt to machines and how much could be produced by automation compared to hand labor. Then, as industry grew, more emphasis was placed on making jobs and tasks fit the people doing the work.

Yet, human factors, or ergonomics, did not gain much public attention until World War II when accidents in military aircraft were blamed on human error. But, in fact, some accidents were caused by the lack of oxygen at high altitudes. As planes were built to fly higher, oxygen equipment had to be developed for pilots. Poor cockpit design also led to some accidents, especially when pilots were not able to quickly and easily determine which buttons or levers to push or pull.

Other members of the armed forces had problems learning to use radar and to operate various types of military equipment. Thus, teams of experts in such fields as anthropometry (the science of human measurement), biomechanics (the study of muscular activity in living creatures), experimental psychology, and engineering had to study physical tolerance and abilities in order to develop guidelines for military equipment design and for training.

Since World War II, the process of designing with people in mind has included studies of work areas and stations in industry and of how much strength and endurance are needed for some jobs. The designs of consumer products, from bicycles and computers to telephones and wristwatches are also a concern of human factors specialists.

The kinds of environmental and product design problems that ergonomists try to solve literally fill volumes. No two design tasks are exactly alike, but all ergonomists use a systems approach to the design process. They view people and the objects people use as parts of total systems. A system might be an operator and a computer, a person riding a bicycle, a pilot using instruments in the cockpit of a plane, a worker inspecting products on an assembly line, or a student trying to read in a library cubicle.

To create a design for such people-place or people-machine systems, human factors specialists follow the same steps used in problem-solving. They identify the problem, or the type of product or place to be designed. Then they research or gather information about possible solutions to the design problem. Finally, they discuss with experts whether a suggested design will work, analyze the possibilities, pick the best idea, then try it out. If the design solution fails, it is back to the drawing board, or the entire idea might be scrapped. If the design succeeds in tests and experiments, production will probably go ahead.

Human factors experiments are sometimes conducted in the field—that is, in a real-life situation. Or research might be carried out in a laboratory, using a variety of testing devices and simulators (imitation or substitute environments).

One of the earliest laboratories for human factors study in private industry was set up in 1960 at the Eastman Kodak plant in Rochester, New York. It all began with Harry L. Davis, an industrial engineer, and Charles I. Miller, a company physician. The two men were concerned about complaints from workers on an assembly line that Davis had designed. Time-study experts had set a standard for the amount of work they thought employees should complete in an hour. But workers said the job was "too hard." They were under a great deal of strain trying to finish tasks according to the time-study standards.

Davis and Miller believed they could find an objective way to measure the demands of the job and whether it was within the capacity of employees to complete. At the same time, they hoped to come up with guidelines that would protect workers from too much stress but still ensure good productivity. Kodak managers agreed that such studies would help not only individual workers but the company as a whole. It made good sense to design jobs so that the largest possible range of employees could do the work. This would reduce stress, possible injuries, and cut absenteeism, while boosting production.

In the early days, the human factors laboratory at Kodak focused on job design only. Under the direction of Dr. Miller, the heart rates and oxygen consumption of workers were measured. This showed how much energy employees used and whether the work was too hard or harmful to their health. The studies also showed how long rest periods should be and what kinds of jobs could be altered to better suit the workers.

Over the years, the range of studies and experiments in the Kodak Human Factors Laboratory has broadened. Now a staff of engineers, psychologists, health professionals, and safety experts evaluate a large variety of jobs and consumer products. As one staff member explained: "Our goal is to fit the job to the person, not the person to the job, and to make sure that Kodak products are easy to handle, operate, and service."

Across the nation, ergonomists or human factors professionals are at work in a number of companies such as IBM (International Business Machines), Xerox, Boeing Aerospace and Airline Companies, McDonnell Douglas, Westinghouse Electric Corporation, American Telegraph and Telephone, Bell Laboratories, Eli Lilly Company, Johnson and Johnson, General Electric,

3M Company, Honeywell, Control Data Corporation, and many others.

Some ergonomists serve as consultants who are called in by businesses and industries. Or they can be found in universities, conducting research and developing scientific methods for studying human capabilities and limitations. Others are in the armed forces and develop military equipment and training programs for military personnel. Or they work for state or local governments on such projects as developing public transportation systems and highway traffic signs that can be clearly understood.

Although an increasing number of professionals are involved in human factors engineering, thousands of companies still do not make use of ergonomic expertise. Unfortunately, many manufacturers and businesses do not see the need for human factors knowledge until a product or workplace causes injuries or is a threat to health and life.

As one ergonomist noted: "Liability lawyers seem to be the people most aware of the impact of human factors engineering." Victims who have been injured or who suffer from ill health due to poorly designed products or hazardous work areas call in attorneys to sue companies for compensation. Yet, many of the facts and data that are revealed in the courtroom could have been used in the beginning stages of design to prevent problems. Taking human factors into account can save billions of dollars in medical payments and design corrections and can also protect the lives of millions of consumers and workers.

ERGONOMICS IN INDUSTRY 2

Work May be Dangerous to Your Health the title of a recent book warns. And a number of articles over the past few years have described job-related problems such as poisoning from chemicals used in the workplace and lung diseases caused by manufacturing or installing such materials as asbestos. And eyestrain and stress have been related to poor placement and design of video displays.

Perhaps the leading occupational injury is back sprain. Working for long periods while bent over and lifting heavy materials without the aid of mechanized equipment ("manual materials handling," as the task is called in industry) cause most of the back injuries suffered by industrial workers. According to the National Safety Council, "Strain and sprain injuries affecting the back account for approximately 35 percent of all worker's compensation hospital and medical claims. This adds up to more than 10 billion dollars a year."

Health and safety managers in industry usually instruct workers on the established and safe way to lift. But proper lifting is not the only safeguard for workers. The work area may need to be designed so that less manual lifting is required. Perhaps platforms for heavy materials can be built or maybe materials can be moved closer to the worker. It might be possible to install mechanical lifting equipment.

Poor design in the workplace (above) can result in an aching back. It would be much better for a worker to use a table that fits the job (below) than to try to adapt to the workstation.

Good lifting techniques include keeping the back straight and lifting with the legs, holding the load close to the body, and *not* twisting the spine. Some of the strain of lifting can be relieved by storing materials off the floor or ground.

Other common work-related injuries affect the shoulders, elbows, wrists, hands, and fingers. Tendonitis is a constant problem for some workers who have to repeat motions over and over. Tendons may be damaged and swell or get sore from pulling a trigger on a paint or power gun all day, or turning a knob constantly, or throwing rejected parts from an assembly line into a bin.

Many workers suffer from carpal tunnel syndrome, an injury to the nerve that runs through a channel, or the carpal tunnel, of the wrist. Usually, the injury comes from bending and twisting the wrist under pressure or force, as when a worker rivets or drills continually. If the wrist nerve is damaged, a person can feel tingling, burning, pain, or numbness. To correct such problems, a worker may have to wear a wrist splint and take time off from the type of work causing the symptoms. In severe cases, surgery might be needed.

Strains, sprains, muscle fatigue, and a variety of injuries may be the result of poor workplace designs in industrial settings. As one British ergonomist, John Hammond, put it: "Many machines found in industry today are nicely styled, with elegant controls. Often, however, they are mounted in such a position that only an operator made in the likeness of a giraffe could effectively handle the controls and at the same time see the work point satisfactorily. And a definite hazard is created when there are unguarded rollers or moving parts..."

When ergonomic principles are applied, the layout of a workstation is designed to fit the capabilities and body sizes of nearly all workers who might use the area. It would not be practical to design a work area to fit every possible worker—for example, the tallest or shortest people. And designing for the average worker would be too limiting. But if designed for the 5th to the 95th percentile person, or 90 percent of all possible workers, a workstation would be suitable for all but 10 percent of the population. The best workstation, however, is one which

ERGONOMICS IN INDUSTRY / 21

is adjustable, allowing workers to change the station to meet their needs.

Many times, job tasks and work areas have to be redesigned to prevent fatigue and strain. In this illustration, it should be obvious which work area (crane cab) is designed for the operator.

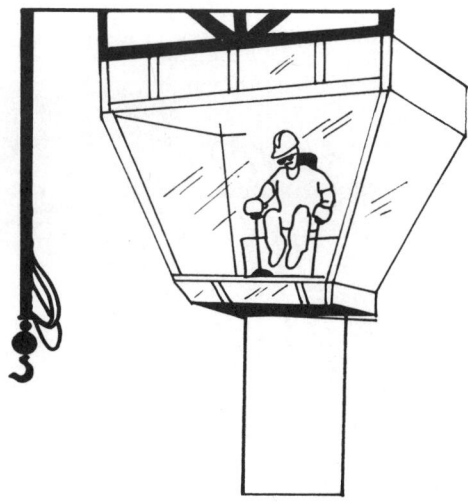

22 / ERGONOMICS IN INDUSTRY

The auto workers union, UAW, has suggested that all auto assembly plants be redesigned so that cars can be turned on their sides, as the illustration below suggests. "This would eliminate the need to work in a pit under the assembly line with arms raised overhead," the union says.

Standing in one spot or holding a tool or workpiece steady for a long time is an example of "static" work. It causes muscles to tire quickly and creates discomfort or pain. Biomechanical studies show that it takes muscles twelve times longer to recover from static work than from tasks performed with dynamic activity, such as walking, in which the muscles tense then relax in a steady rhythm.

Poorly designed tools also create health and safety hazards for workers. Although most hand tools have developed through trial and error, new tools do appear on occasion, and many older tools have been modified. Ergonomists recommend that hand tools perform the functions for which they were designed. They should be well-balanced and easy to hold, producing little fatigue.

ERGONOMICS IN INDUSTRY / 23

Workers should be able to operate hand tools without bending their wrists, and use power tools, with a minimum of vibration, whenever and wherever possible.

Here is an example of both "static" and "dynamic" work. While hammering, a worker could use a clamp to hold the board and relieve the muscle tenseness from "static" work.

In many workplaces today, robots are using tools. These automated workers are "on the march," as a recent issue of *U.S. News and World Report* put it. The magazine also predicts that " 'intelligent robots' with *visual* and *tactile* senses will be widely used in assembly work" in the years ahead.

Robots can save workers from dull, exhausting, or hazardous jobs. For example, in the General Electric plant in Erie, Pennsylvania, "praying mantis"-type robots pick up tools and drop them into milling machines that make motor frames. Robots repeat such tasks over and over with precise timing, hour after hour. It would be impossible for people to work as efficiently at such a steady pace.

Some robots can carry heavy objects—from 132 to 210 pounds (60 to 95 kg)—that would be unsafe for many humans to lift. In auto manufacturing, robots do tedious jobs such as spot welding or spray painting. Robots are used in dangerous police work. In fact, RMI-3, a robot in the New York City police department, won a "Cop-of-the-Month" award in April 1984. Robots are also used for testing equipment in hostile conditions such as nuclear or battlefield environments.

In spite of the safety and health benefits of using robots, many workers resent these automated helpers. They often contribute to higher unemployment in industry. Yet, there is really little chance that robots will work without human operators or monitors.

In future years there will be a "tremendous need" for workers who can maintain and repair robotic machinery, says Dean Eavey, head of a robotic training program at Vincennes (Indiana) University. Increasingly, ergonomists will be concerned with the design of robotic training programs as well as the design of the robots themselves. If human factors are considered, robots should be safe and easy to maintain and be aids for the operators.

Whether a workplace is automated or not, workers need the protection of warning signals such as sirens and buzzers and lights. Various control devices are needed also. Control devices can be buttons, levers, knobs, handles, pedals, wheels, cranks—anything that puts information into a system, whether it is a simple mechanism or a complex machine. When determining the proper controls for machines or equipment, ergonomists consider such factors as size, shape, and texture; where controls will be located and how far apart; how much force is needed to properly and safely activate a control; and which direction and how far controls should move.

Identifying controls may not be critical for some tasks, such as finding the right switch or button to turn an office computer or a vacuum cleaner on or off. But in a number of instances, people have to be able to correctly identify control devices in a matter of seconds to avoid disastrous results.

One human factors study found that engineers of some diesel locomotives mistakenly grasped the wrong controls to turn off signal lights. Instead, they shut off the fuel pumps, stopping the train engines. Drivers of some older model cars have had the frightening experience of accidentally pressing down on a gas pedal when they intended to stop. In such cases, the gas and brake pedal were not only similar but placed too close together.

During World War II, the U.S. Air Force conducted an extensive study of rapid identification of controls. Researchers found that over 400 aircraft accidents in a twenty-two-month period were caused by pilots who were confused by landing gear and flap controls that were too much alike. Since that time, the Air Force, along with the rest of the military, has mandated that human factors be considered in the design of controls and other aspects of equipment, machines, vehicles, and aircraft.

Shape-coded knobs are a must in aircraft control devices. The U.S. Air Force has designed three different types of knobs for specific purposes, which a pilot can identify simply by touch. Some knobs are also shape-coded so that they symbolize their functions. For instance, the control knob for the landing flap looks like a miniature flap and a wheel-type knob represents the landing gear.

Location is another important consideration for aircraft controls. At Charles Mauro Associates in New York City, a team of human factors engineers found that a major airline crash of a Convair 580 was probably due to a poorly located push-button control. The Convair crash, which took twenty-seven lives, happened at Chicago O'Hare International Airport in 1968, but the causes were not known until years later. Relatives of the pilots who died in the crash sued the manufacturer of the automatic pilot system installed in the Convair. In 1979, the Mauro consulting firm was called in to investigate.

Human factors engineers at Mauro spent a year studying possible causes of the system failure. During the investigation, the engineers made a replica of part of the cockpit. They discovered that, because of the location of one of the control buttons, a pilot may have accidentally pressed it with his elbow. If pressed, the control would have given false information on a display panel, possibly causing the pilot to make the wrong maneuvers and crash the plane.

Of course, no one could prove absolutely that the pilot's elbow accidentally pressed against the control button. But the probability was strong enough to convince the court, which awarded the victims' families $3.5 million in compensation.

In an entirely different work situation, the oil field drilling services industry, workers have to operate so many controls at such a great pace that they are under great stress. Blenders are used during hydraulic fracturing, a process in which a dense mixture of water, sand, and assorted chemicals are pumped down a well at very high pressure, creating fractures in the oil or gas zone.

During this process, many pieces of equipment—jet turbines, controlling vans, chemical trucks, and mechanical blenders—are assembled near the well head. A maze of high pressure pipes and pumps covers one or two acres. As many as forty operators may be needed to control the equipment. Since the noise level is very high, workers must communicate through hand signals and radio. As Stephen Casey, senior scientist and ergonomist at a consulting firm in Santa Barbara, California, points out: "Engineers have designed highly reliable equipment for the oil field, but almost all controlling functions are left to the operator. The result is frequently a textbook example of operator overload," and performance falls off.

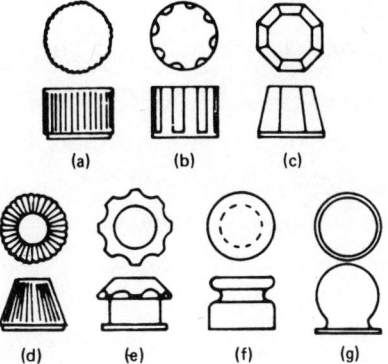

Multiple Rotation Knobs: These knobs are for use on controls that require twirling or spinning, for which the adjustment range is one full turn or more, and for which knob position is not a critical item of information.

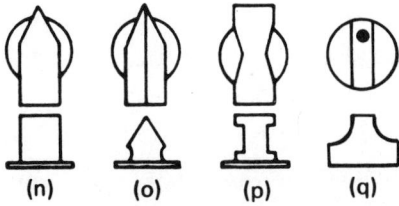

Detent Positioning Knobs: These knobs are for use on discrete setting controls, for which position is critical.

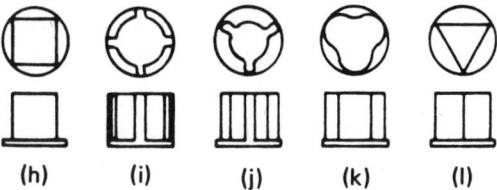

Fractional Rotation Knobs: These knobs are for use on controls that do not require twirling or spinning, for which the adjustment range is less than one full turn; and for which the knob position is not a critical item of information.

To illustrate his point, Casey, who has designed oil-field equipment, described how the operator of a blender system must work. "The blender operator is the keystone in the system," Casey explained. Working from a console that controls treatment lines, the operator has sole responsibility for mixing sand, water, and chemicals in exact proportions. The materials must enter a mixing tank at the same rate as the output which can be as high as forty barrels per minute.

"Not surprisingly, the typical blender operator must be very alert throughout the course of an eight-hour job," Casey says. However, he adds that because of microprocessors, "the industry as a whole is now capable of automating many functions traditionally performed by operators," a process recommended by ergonomists, including Casey. He believes more attention will be paid to human factors in the design of oil-field service systems in the years ahead. Future operators should be able to monitor automated systems rather than be responsible for controlling so many service functions, thus reducing the possibility of errors.

Yet, some highly automated systems can be hazards in themselves. This is especially true when those who monitor the systems cannot easily identify controls or intervene in emergencies. Evidence of this surfaced during the 1979 investigation of the near meltdown at the Three Mile Island (TMI) nuclear power plant in Pennsylvania.

Operators in the TMI control room had to monitor hundreds of signal lights and meters and be able to distinguish among and use hundreds more control levers. When the cooling system failed at the power plant (which led to the shutdown of a reactor and release of radioactivity), workers did not detect the problem right away. For one thing, they could not see one of the warning lights that signaled a feedwater valve was closed. A cardboard maintenance tag had been left dangling on the control panel, covering the signal. In addition, the Nuclear Regulatory Commission (NRC) later found that "poor panel layout" in the

Inside the control room of a nuclear power plant, observers study some of the hundreds of signals that must be monitored by operators.

control room made it difficult to spot problems in the power system.

A typical control room had a panel 8 feet high and up to 100 feet long. Rows and rows of levers that looked exactly alike lined the panel. In some cases, meters were labeled from left to right, but meter switches were set up and labeled in just the opposite manner, from right to left.

In other instances, control handles had to be adjusted on one side of the room as displays across the room showed what was happening. Along the top of the control panel, signals lit up to show any abnormal conditions. Other signals—totaling over 3,000—indicated when electric controls or pumps were on or off.

During an emergency, such as a loss of coolant, hundreds of signal lights could go on and off at the same time. And a red signal light would not necessarily send a familiar message such as "danger" or "stop." Instead, it might simply mean a switch or valve was closed.

With so many and such confusing controls and signals, it was a near miracle that TMI operators could take any emergency action, some human factors experts have reported. Yet, it would have been possible to build a control room according to ergonomic principles. As Professor Thomas Sheridan of the Massachusetts Institute of Technology wrote in *Technology Review*, "A wealth of human factors engineering literature is available... to provide detailed design guidance for individual displays, controls, and operating procedures." But Sheridan went on to warn that "Concern for human factors and human errors in nuclear power plants should not be limited to the control room." He pointed up the need for integrating human factors at every stage from the design of the plant to manufacturing and testing of equipment to maintenance and repair.

Since the accident at TMI, the Nuclear Regulatory Commission has set up a Division of Human Factors Safety. The group is responsible for designing programs to train and license nuclear plant personnel and to establish emergency procedures. The division has also set up hundreds of guidelines—many based on long-established ergonomic principles—for control room design. The guidelines cover the layout of the work space, proper lighting, signal and communication systems, visual displays, color coding, and the types of controls that should be used.

New control panels or redesigned controls for nuclear plants are being developed around new computer systems that, it is hoped, can be easily understood. But, as Sheridan cautions, the development of computer technology may be "well ahead of our understanding of how humans might best use such technology and the new kinds of errors that might occur. This clearly is an area of great promise, but also one of relative ignorance. There is much catching up to do."

"HUMANIZING" THE OFFICE 3

Manufacturers and advertisers of computers often refer to their products as "user friendly." But a large number of the seven million or more Americans who use computers in their daily work would dispute that claim. Since the mid-1970s, when computers became common equipment in many workplaces—from air traffic control towers to offices—operators have increasingly grumbled and growled about the decidedly "unfriendly" manners of their electronic equipment.

In 1980, clerks at Blue Shield (health insurance) of California went on strike to protest, among other things, the eyestrain caused by poor lighting and glare from video display terminals (VDTs), or computer screens. Members of the American Newspaper Guild, who were the first workers to use VDTs on the job, have long complained about eye fatigue, body aches and pains, dizziness, nausea, and headaches from computer screens. According to Charles A. Perlik, president of the Guild, 1100 guild members were surveyed over a two-year period as part of a study conducted by the Mt. Sinai School of Medicine in New York City. The report of the study, released in October 1983, showed that computer terminal users lost more time from work than nonusers, suffered deteriorated vision, and had eye irritation.

Yet, an earlier study by the Association of Ophthalmologists of Quebec, Canada, found that working continually over a five-year period on VDTs caused "no harmful effects on the occular and visual systems" of operators. In 1983, the research panel of the National Academy of Sciences agreed, concluding that VDTs in themselves did not cause any more eye discomfort than other types of close work. The research panel, which included several ergonomists, said its study showed eyestrain "may relate to how the VDT is designed and where it is positioned in the workplace and the nature of the job that requires a worker to spend many hours in front of the screen."

Another concern of VDT users is possible radiation effects from video displays. The National Institute for Occupational Safety and Health (NIOSH) found that radiation exposure for VDT users is well below the safety limits set for such hazards. And, at times, radiation cannot even be detected by survey instruments. Other research groups—the Bureau of Radiological Health, Duke University, and AT&T Bell Laboratories—have also concluded that VDTs pose no radiation risk to those who use them. Nevertheless, the Newspaper Guild, 9 to 5—a national organization of female office workers—and other labor groups want further investigations because of a large number of miscarriages among computer operators and birth defects among children of VDT users.

While there is still much debate about some effects of VDTs, there is general agreement that worker problems in the electronic office are due to multiple causes. Many times computers are placed in unsuitable workplaces, such as areas where ventilation is poor. Poor air circulation can be a special problem for VDT users because of the heat from electronic equipment.

Improper lighting can be damaging to the health of office workers. Designers recommend that subdued indirect lighting rather than intense overhead lighting be used. Individual lights at workstations, especially where computer operators must spend long hours, can prevent eyestrain. To reduce glare, VDT screens should be at right angles to windows and other light sources, and dull-finish materials on desks and other equipment should be used to prevent reflection.

High noise levels in open offices can cause stress. But noise can be cut by using sound-absorbing materials on partitions, ceilings, floors and furniture. "White noise" also blocks sound. This is a low-level background noise, like the sound of rushing water, that is electronically produced. It helps mask other sounds that would be disturbing or stressful.

Uncomfortable desks and chairs may be some of the most hazardous components of an office. Furniture that does not fit the user causes muscular aches and pains, tension, and a variety of other ailments. At computer workstations, "ergonomic furniture" can make a difference in the level of productivity.

The U.S. Department of Health and Human Services measured productivity of workers in "best designed" and "worst designed" computer stations. Workers in the best designed stations increased their output by 25 percent. Another study, carried out by Louis Harris Associates for a major office furniture manufacturer, showed that comfortable furniture and environments were important factors in employee satisfaction, which in turn boosted performance.

The total office environment should enhance the quality of work life, human factors specialists believe. According to David Armstrong, director of Facility Management Institute, a consulting firm that concentrates on office environments, "The chemistry of the workplace is strongly affected by the physical surroundings—where people work in relation to other people, how functional their own workspace is, and the general tone of the place."

Yet, how is total office comfort determined? What makes a good desk or chair? Who decides the arrangement of an ergonomic workplace?

After some seventy years designing, manufacturing and selling office furniture, Steelcase, Inc., believes it has some of the answers. In 1983 the company opened new corporate headquarters in Grand Rapids, Michigan, and the building has been described as "a state-of-the-art office environment." The five-story tiered structure was literally designed from the inside out. Its shape, size, and form evolved around the needs of the people who work within. Employees helped determine the arrangement and content of individual workstations and where various departments would be located. In the open-plan offices, furniture fits job functions, not the status of the workers.

Flexibility is the key to a well-designed office, experts have found. People should be able to adjust their desks, chairs, machines, lighting, and partitions to fit individual needs. Ideally, a computer desk, for example, should have a split top so that the VDT can be placed on one half and the keyboard on the other. If the desk has been designed for physical differences, the height can be changed up or down, allowing the worker to position the monitor and keyboard to suit personal preferences.

Human factors engineers and occupational health experts know that poorly designed chairs and seating can be a hazard in almost any workplace, but chairs that are misfits and uncomfortable can be especially harmful in the office. As Jack Hockenberry, human factors engineer specializing in office products put it: "Many so-called comfortable desk chairs have not been designed so that the user can avoid discomfort during long-term fixed-position sitting tasks . . . . As the discomfort slowly increases, the seated person is distracted from work tasks because attention is drawn to the pain caused by the chair. The error rate goes up, and productivity is greatly reduced. The body support product or operator seat is a direct contributor to the alertness level of the user. The scientific application of anthropometric data to the design of office seating has demonstrated that

"HUMANIZING" THE OFFICE / 35

Since the late 1960s, Steelcase designers and engineers have been studying the needs of people who use electronic office equipment. They have found these factors most critical: Keyboard height, eye-to-screen distance, viewing angle, hand-to-keyboard distance, seat height, and back support. The Steelcase computer workstation, as shown in this diagram, has (A) a seat back and (B) seat that tilts; (C) an adjustment for seat height; (D) a range of 18" to 22" from the screen for normal viewing; (E,F,G) vertical angle and height adjustments for the screen to minimize eye, neck and shoulder fatigue; (H) keyboard to screen adjustments; (I,J) keyboard height and angle adjustments.

alertness and productivity are improved when seat and back contours are designed to fit the user."

What kind of anthropometric data—scientific measurements of the body—do ergonomists use? Hundreds of different body features have been measured for various age, sex, occupation, and ethnic groups, and standard anthropometric tables have been published to assist designers. The data may indicate, for example, the sitting height for 5, 50, or 95 percent of all adult females or males.

Based on such data, some basic principles have been developed for the design of chairs and other types of seating. According to some of these principles, a properly designed chair should have an adjustable seat height so that the feet can rest squarely on the floor. A seat should be parallel to the floor with a scrolled or "waterfall" front to reduce behind-the-knees pressure that causes circulation problems. A backrest should fit the natural curvature of the spine and support the vertebrae that carry most of the upper body weight. And armrests should adjust for best support.

To make sure that such human factors are considered for automated offices, NIOSH has set up guidelines that stress flexibility in the design of furniture and also suggest standards for electronic equipment. A number of states—Connecticut, Illinois, Maine, Massachusetts, New York, Oregon, and Washington—are considering laws that require companies to follow the NIOSH guidelines.

Some ergonomists, however, are concerned that "mandatory standards" for VDTs and other computer equipment could be "premature." Since the technology is changing so rapidly, rigid standards might stifle improvements. For example, a number of European studies have been conducted to determine the most effective color for video monitor screens. In one German study, operators preferred green, yellow, and amber screens and their overall performance with a yellow screen and amber filter was four times greater than with a black and white display. Some European manufacturers are using green, amber, and yellow in their video monitors, and a few American manufacturers are offering a choice of video display colors. But there is still no universal agreement on the best color combinations to use.

Along with an emphasis on well-designed equipment and furniture, ergonomists are also stressing the importance of easy-to-understand computer programs. The software and "documentation," or instructions, can make the difference in whether or not an operator decides a computer is "user friendly." When program instructions are difficult to follow and use, the result is frustration, anxiety, and stress, which, over a period of time, can lead to health problems.

Another human factors consideration in the automated office is the nature of the work itself. As Karen Nussbaum, executive director of 9 to 5, wrote in *Working Woman*, "Office workers have gained few of the potential benefits of computer technology. Most often, clerical workers at terminals find they end up with only one task to perform, which they must repeat again and again at a pace and according to standards imposed by the machine . . . . In many cases, automation has destroyed the very aspects of work that most clerical workers enjoy—variety within jobs, relations with co-workers and control over work style."

Computer operators may feel they are only parts or extensions of machines. In some companies the machines are even "watching" the users. Computers monitor the input and output of their operators and, in some cases, workers' wages are adjusted (as with piecework in factories) according to the amount of data entered into the computers. Yet, computer operators seldom receive pay raises that match the higher productivity of automated machines.

Getting workers involved in how they work—planning and

Adult Male

Selected Anthropometrics Features 1960-62 National Health Survey of the United States Public Health Service.

Body Feature	Male, Percentile		
	5th	50th	95th*
A Height, inches	63.6	68.3	72.8
B Sitting height			
– Erect	33.2	35.7	38.0
– Normal	31.6	34.1	36.6
C Knee height	19.3	21.4	23.4
D Popliteal height	15.5	17.3	19.3
E Elbow-rest height	7.4	9.5	11.6
F Thigh-clearance	4.3	5.7	6.9
G Buttock-knee length	21.3	23.3	25.2
H Buttock-popliteal length	17.3	19.5	21.6
I Elbow-to-elbow breadth	13.7	16.5	19.9
J Seat breadth	12.2	14.0	15.9
– Weight, pounds	126	166	217

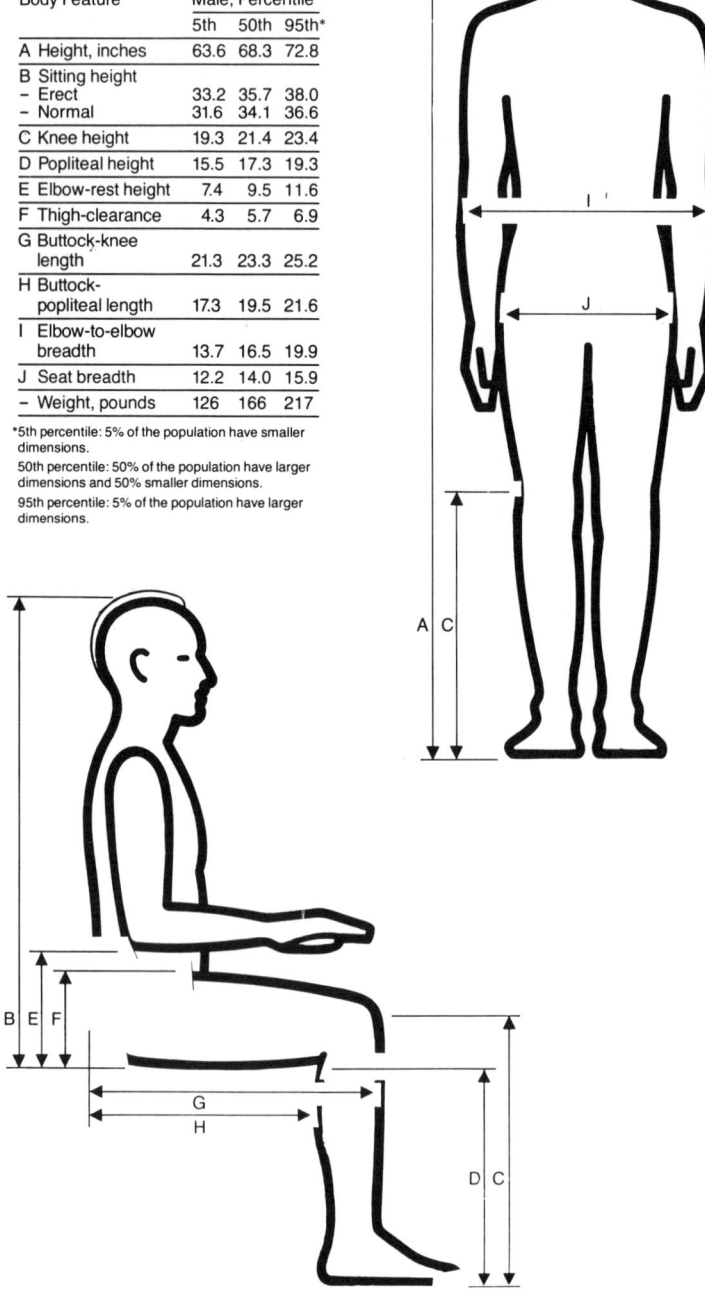

*5th percentile: 5% of the population have smaller dimensions.

50th percentile: 50% of the population have larger dimensions and 50% smaller dimensions.

95th percentile: 5% of the population have larger dimensions.

Adult Female

Selected Anthropometrics Features
1960-62 National Health Survey of the
United States Public Health Service.

Body Feature	Female, Percentile		
	5th	50th	95th*
A Height, inches	59.0	62.9	67.1
B Sitting height			
- Erect	30.9	33.4	35.7
- Normal	29.6	32.3	34.7
C Knee height	17.9	19.6	21.5
D Popliteal height	14.0	15.7	17.5
E Elbow-rest height	7.1	9.2	11.0
F Thigh-clearance	4.1	5.4	6.9
G Buttock-knee length	20.4	22.4	24.6
H Buttock-popliteal length	17.0	18.9	21.0
I Elbow-to-elbow breadth	12.3	15.1	19.3
J Seat breadth	12.3	14.3	17.1
- Weight, pounds	104	137	199

*5th percentile: 5% of the population have smaller dimensions.

50th percentile: 50% of the population have larger dimensions and 50% smaller dimensions.

95th percentile: 5% of the population have larger dimensions.

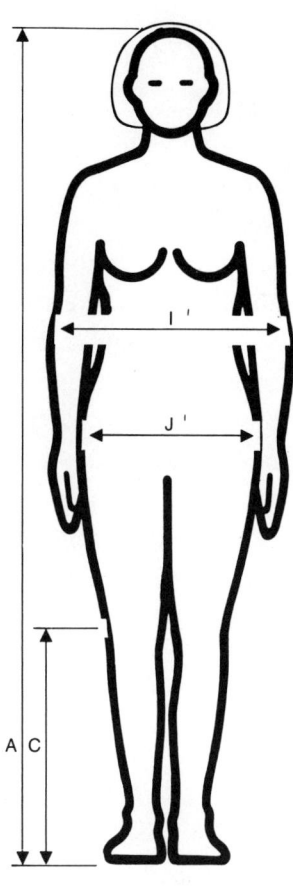

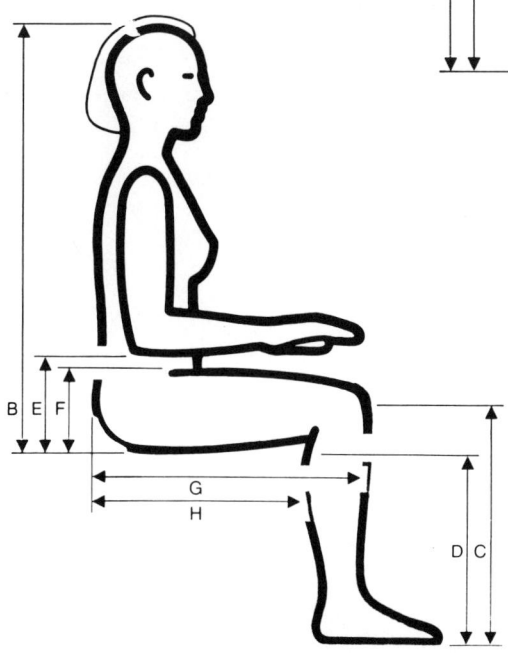

organizing tasks—is one more important human factors consideration in automated offices. People often resist change and fear automation. They feel threatened because they must learn new ways to work or face the loss of their jobs. As a result, workers resist new technology, stress on them builds, and, with increased absenteeism and errors, performance declines.

Increasingly, office workers are asking for a role in designing their job tasks. Clerical jobs, for example, could be designed so that workers would not have to perform the same tasks repeatedly but could take on a variety of related responsibilities. Many studies show that workers in both offices and in manufacturing jobs improve their productivity when they take part in decision-making and are treated as "valued partners" rather than mere cogs in the machinery.

CREATING CLEAR VISUAL MESSAGES 4

Everyone is confronted with a variety of visual messages throughout the day: signs (with both words and symbols), labels, instruction manuals, meters and dials, coded keys and buttons, signal lights. Some visual messages are intended to protect public health and safety. Others provide information. They identify places and things, give directions, or label materials.

Paying attention to human factors principles can make this transfer of information easier. Ergonomists sometimes advise designers on the types of graphics or of written instructions that will best convey messages. Of course, some visual messages already communicate well; they are clearly presented and are in standard, familiar forms. A green "Mr. Yuk" label quickly warns young children of the danger of poisonous materials.

The happy/sad face is a symbol that almost everyone understands:

42 / CREATING CLEAR VISUAL MESSAGES

The slash across a simple, bold symbol is another common graphic device that puts across such traffic messages as "no trucks," "no bicycles," "no U turn," "no left turn."

Other familiar symbols on traffic signs quickly provide information such as those that show deer or cattle crossings or that there are restaurants, gas stations, or other service places available just off the exit road. The signs that mark parking places and restrooms for the handicapped hardly need words to underscore their meaning:

But what about these signs?

BLIND RR XING
DISABLED CARS REQUIRED TO PULL OFF ROADWAY
PLEASE WAIT FOR HOSTESS TO BE SEATED

No one really expects a blind railroad to grope its way across a street or a disabled car to pull itself off the road. And few people entering a restaurant would stick around waiting for a hostess to seat herself before ushering customers to their tables. The point is, even though most people would understand these messages, the choice and arrangement of words are misleading. They do not say what they are intended to mean.

Some signs, labels, instructions, notices, or other visual messages can be truly confusing, as well as amusing. Here is one found on a small office machine: "Disconnect Power Cord From Branch Circuit Supply Outlet (before removing cover)." Office workers should be excused if they fail to understand such instructions. The directions could have simply said: "Unplug machine before removing cover."

On an Army base, this notice was posted: "No security regulations shall be distributed to personnel that are out of date." It is possible the people working on the base were "out of date," but the phrase actually referred to the security regulations. Maybe a simple command would have been better: "Don't Distribute Out-of-Date Security Regulations."

Even more confusing was a government official's report on what he found while inspecting a regional office: "To the extent rendered possible by the limited amount of time allotted to the investigation of this area, it appears generally that assignments and work schedules were, in large measure, being carried out in accordance with desirable standards." The official could have written: "As far as I could tell in my short visit, the regional office is getting the work done fairly well."

44 / CREATING CLEAR VISUAL MESSAGES

Then there was the notice tacked to a bulletin board in the teachers' lounge of a high school: "All appropriate personnel shall endeavor to give full compliance to the regulations herein promulgated." Perhaps, if teachers read the notice, they got the idea that all school staff should try to follow the rules!

As the examples have shown, simplicity is the key to well-designed messages. The kind of typeface used for signs, labels, and other printed forms also makes a difference in whether or not a message can be easily understood. Plain typefaces make printed copy more readable.

A plain typeface like this, for example, is easy to read.

BUT USING ALL UPPER-CASE LETTERS IN LONG PARAGRAPHS CAN MAKE DIFFICULT READING.

However, capital letters in HEADINGS for paragraphs and in short sentences can alert readers to important messages.

A person with good eyesight and plenty of time and patience can decipher a message in this decorative typeface, but fancy or unusual type is not recommended for quick, easy reading.

Many printed materials, signs, and labels have to convey messages to people with varied abilities and from diverse backgrounds, so those human aspects should be considered in the design process. Ergonomists also recommend that designers understand where visuals are going to be used and under what conditions. For example, in the two photos above, the letters on the door (left) seem to indicate who should use the lavatory. But a person in a hurry could easily misread the word. From this angle in the hallway, all of the letters are not visible on the recessed door. Barging inside, men have missed the full word (right). A sign for the lavatory could have been placed on the wall or be extended beside the door to prevent confusion—and embarrassment!

46 / CREATING CLEAR VISUAL MESSAGES

If signs or labels have to be read from a great distance, the height, width, and spacing of letters and numbers are of special concern. In some workplaces, paper labels and signs may need a protective coating to prevent their being damaged by dirt or chemicals.

Signs and labels that convey warnings or instructions need to be in plain view and clearly presented. According to human factors specialist Frank Fowler, "Most industries relegate the writing of instructional manuals and warnings to a technical writer or the low engineer on the totem pole," making it more likely that an error may be made or that someone will not understand the written material. Fowler testified in court on behalf of a worker who had contracted a neurological disease because of welding fumes. According to Fowler, "the warnings printed on the welding rod boxes were inadequate, even though they complied with the standard of the industry. By any measure of human factors principles, the warnings were totally inadequate. The plaintiff won to the tune of $500,000 on that basis."

Color-coding may also help put across clear visual messages. Americans learn at an early age what red, green, and yellow mean. Red stands for danger and means "stop" in traffic. Green indicates "go" or is a forward/on position for some equipment and machinery. Yellow and bright orange caution most people.

Yet, what if colors are not always used as expected? Suppose warnings were printed in lavender or pink letters? Few people would really pay attention because those colors are not usually associated with risk or danger.

In some workplaces, a red signal light might show that a machine is "on" and a green light might mean there is a problem. Such a situation could easily confuse a new worker who is "used to" just the opposite meanings for red and green.

CREATING CLEAR VISUAL MESSAGES / 47

When visual displays—whether signs or signals, buttons or dials—operate contrary to or differently from what most people expect, the result is usually frustration and an increased chance for errors. Look at the layout for numbers on a push-button telephone. Most people expect the numbers to start at the top and go across several rows in this manner:

```
    1   2   3

    4   5   6

    7   8   9

    *   0   #
```

But some new phone designs have panels numbered directly across in a single row:

1   2   3   4   5   6   7   8   9   0

It might be confusing and difficult for a user to move back and forth across one row to call a long number. Both designs would have to be tested for ease and simplicity of use before ergonomists could determine objectively which layout best served users.

The numerals on calculator keys are in a different order, starting on the bottom row and increase to the top in this manner:

```
    7   8   9

    4   5   6

    1   2   3

        0
```

48 / CREATING CLEAR VISUAL MESSAGES

The layouts for the calculator and the telephone both work very well even though they do not match, but many ergonomists believe a uniform design for use on both machines would reduce errors and frustrations. But which layout should serve as the standard? Manufacturers of calculators would argue for their design while telephone companies would favor their pushbutton order. There seems little chance that one design will be accepted over another in the near future.

Ergonomists also urge some uniformity in the layouts and markings for dials on stoves and other appliances.

"In many cases continuous controls violate the human factors guidelines that relate 'on' and 'increase' to up, right, or forward motions," says Jack Laveson, chairman of the Consumer Products Group of the Human Factors Society. Laveson points out that surface burner controls on some electric stoves move clockwise but the settings are marked from high and rotate to low. Yet, most people expect a low to high rotation—an increase in temperature with a clockwise or rightward motion, as when a flame increases on a gas stove or the temperature goes higher when a thermostat dial is turned clockwise or a lever is pushed to the right.

Another common problem on stoves is the layout of the burners, which is a type of visual display. On an unfamiliar stove, a person usually has to puzzle over which knob or dial controls which surface burner. If the controls are not marked clearly it takes a number of trials and errors to learn the layout. As these diagrams show, the surface burners and the knobs that control them could be set up in a variety of ways:

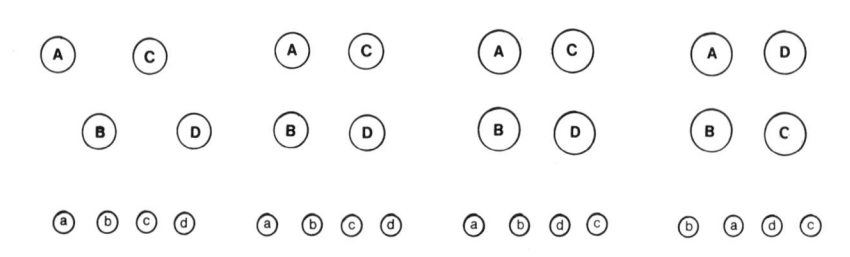

A well-known study by Alphonse Chapanis, a human factors authority in the field of applied psychology, found that stove burners should be offset, as in the first diagram, so that controls clearly relate. Out of 1200 trials with the four different types of arrangements shown, the Chapanis test group made no errors with the offset display.

The order of meters and other measurement devices in workplaces can have an effect on how well a person reads and understands the information presented there. When such layouts are being designed, human factors engineers consider a variety of principles. One is locating meters and other displays in accordance with the order and frequency that they will be used. Another is placing identification and instruction labels close to the instruments they relate to. Such principles may seem like nothing more than common sense, but they are not always followed. A designer may be more concerned with making machines, equipment, and tools work than how people interact with them.

Almost everyone depends on some type of visual signaling device for information during the day. Maybe a strobe light flashes on a school bus to warn drivers in other vehicles to be cautious. Yellow lights atop barricades flash to warn of road and bridge repairs underway. Red lights flashing at railroad crossings stop traffic for oncoming trains. Flashing blue lights on vehicles tell other drivers to make way for volunteer fire fighters. Brake and turn-signal lights on vehicles, light reflectors on bikes, luminous clothing and tape, and lights showing that machinery or equipment is operating are other examples of visual signaling devices in use daily.

Most of these signals have been designed with safety in mind and have helped to reduce accidents. But what happens when signs and signals "fight" for attention, as in the scenes on the next page?

It seems clear that human factors get short shrift along many business streets. Although local governments regulate the number, size, location and use of signs and flashing signals, few cities and towns have enough personnel to enforce all sign regulations. And some cities simply do not place a high priority on cleaning up the "visual environment." Perhaps, some urban areas need an "ergonomics awareness campaign" calling attention to the fact that visual pollution creates confusion!

DESIGNING FOR CONSUMERS 5

A decorator phone may have a great "antique look," but a user soon gets tired holding the separate earpiece in one hand while grasping the base of the phone in the other trying to get the mouthpiece at just the right angle to talk into it.

A camper van seems to have every kind of convenience and comfort for the road, but it takes a Sampson-like strength to push up the "pop top."

A calculator is just the right size for a shirt pocket or purse, but the keys are so small and close together that even the most careful user makes mistakes.

A sleek sofa fits perfectly in the family room, but the seat area is too deep for short people, who have to let their feet dangle off the floor when up against the back support.

A digital clock in the family car keeps perfect time, but it is almost impossible to read when bright sunlight creates a glare on the glass covering and distorts the numbers.

Almost any product or mass-produced item has limitations for some consumers. But many everyday objects could be better designed so that they fit the majority of users. Often engineers give priority to the way a product works while designers are concerned about the way a product looks, but neither adequately considers how the consumer is to use it.

Henry Dreyfuss, who was one of America's first and best-known industrial designers, reminded fellow professionals to "bear in mind that the object being worked on is going to be ridden in, sat upon, looked at, talked into, activated, operated, or in some other way used by people individually or en masse."

Dreyfuss was a pioneer in the study of human factors, and he and his firm designed dozens of popular products from the convenient Trimline telephone to the handy Thermos bottle. The products have been well received by the public because the designers never forgot that the items were meant to be "used by people."

Industrial design and human factors "should be close natural allies," says John Kreifeldt, Ph.D., professor of Engineering Design at Tufts University. Kreifeldt believes the design of consumer products should be integrated not only with human factors but also with manufacturing, marketing, advertising, and product liability areas. The entire process, from the idea for a product to ultimate use, should be unified, Kreifeldt says.

How do human factors specialists fit into this process? Under *ideal* conditions, they are in on the beginning stages, helping to develop ideas for the design of products that range from ballpoint pens to automatic typewriters. Then human factors input can aid in the development and testing stages, providing data on how consumers are expected to interact with the product and how a product might be adapted in shape or operation to suit users better.

Requiring complex machines to fit the people who use them is one thing, but is it really necessary to worry about the design of a simple product such as a pencil or pen? There is little doubt what that answer would be if the pen point retracted while writing or the pencil broke in half with little or no pressure on it.

"Any item a person uses involves human factors, and they should all be looked at," said Nick Simonelli, human factors specialist at 3M Company in St. Paul, Minnesota. Simonelli and

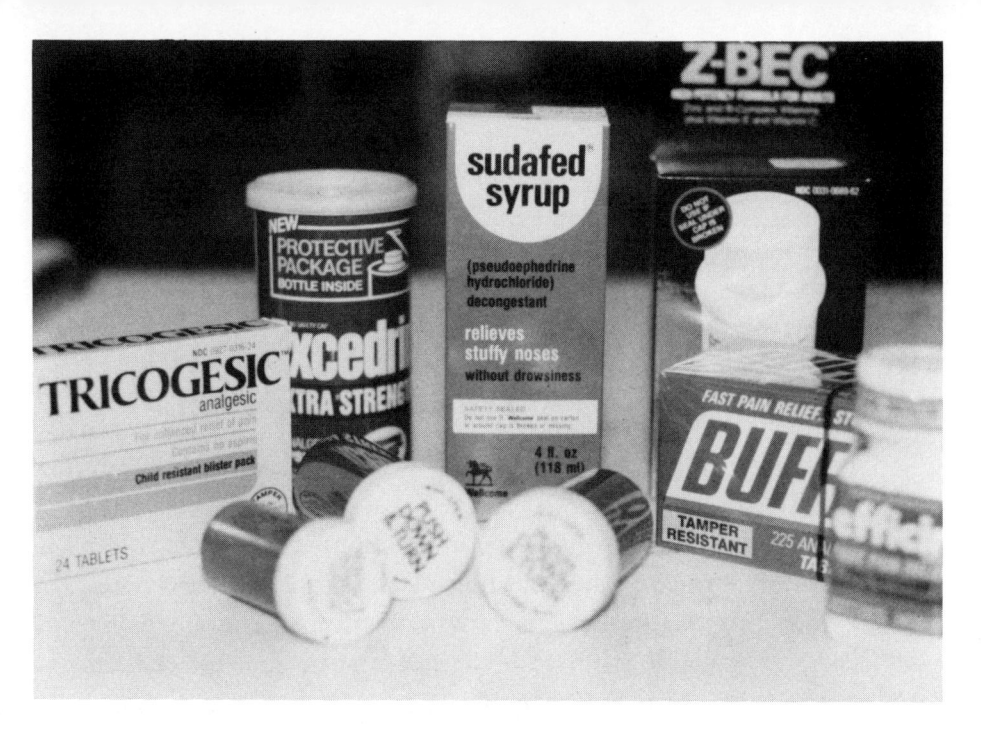

Tamper-proof packages for vitamins and over-the-counter medications protect the health and safety of users, but the elderly may need help to get the tightly sealed boxes and bottles open.

---

four other researchers at 3M would like to be involved in the design and development of most if not all of the hundreds of products manufactured at 3M plants. As generalists, the human factors researchers do work that overlaps the interests of other departments such as engineering and marketing research. "But the difference between human factors and other fields is that we take a systematic look at all possible uses and users [of 3M products] and any and all errors that can be made," Simonelli explained.

At 3M, the human factors staff began with work on copying machines, but since 1981 they have been a resource group for other divisions in the company including Packaging Systems, Medical Products, and Commercial Tape Products. To find trouble spots, 3M's researchers have tested such aspects as whether a left-handed person can easily use a tape dispenser or what the best speed should be for the automatic paper intake on a copier.

54 / DESIGNING FOR CONSUMERS

At many corporations, human factors principles are becoming more and more a part of making products as safe, simple, and efficient to use as possible. At the developmental stage of a product, ergonomists also now consider whether a product fills a real need in the most effective manner.

Such challenges faced human factors experts at AT&T Bell Laboratories during the process of designing standard instructions for public telephones. Since there are so many variations in the operation and use of public phones nationwide, users have had problems trying to figure out how to place calls. At some public phones, for example, callers have to deposit coins before dialing and the money is returned if the call is not completed or there is no charge. At other public phones, a person places a call and waits for computerized instructions (a synthesized voice) on how many coins to deposit before a connection is made.

With regard to long-distance phoning, there are Charge-A-Call telephones that accept no coins at all—callers charge their calls to credit card numbers. From some public phones, users "direct dial" long distance numbers and then pay for the call. But at other phones, the operator asks for the number, places the call, and collects the proper coins.

With variations for using credit card numbers, making emergency calls, getting directory assistance or repairs, even experienced users have to study instructions in order to place calls from public telephones. But until standardized instruction cards came along, the instructions also varied—in layout, color schemes, type styles and where they were located on the phone.

To begin the process of standardizing Bell's instruction cards, human factors specialists interviewed hundreds of public telephone users in order to identify what information was needed and how instructions were used. This information from users along with the Bell System's ideas for a standard set of instructions and a standard layout were then turned over to New York design firm Henry Dreyfuss Associates (named for the late designer). The firm's designs were evaluated in laboratory settings. Paid participants answered questions about the use of the instruction cards. The cards were also tested in "real world" sites such as shopping centers and a major airport. Finally a design was approved. Bell now has installed the standard instruction cards in the public telephones of many Bell systems, and since the breakup of the large corporation has encouraged independent companies to use the new design. Researchers expect their future studies will show that the redesigned instructions have made public telephones easier to use.

Coin needed for all calls   1.   2.

Local calls Deposit 10¢ before dialing
Long Distance Dial 0
   Operator will handle all Long Distance calls

SOS dial 0 for Emergency help

KEENE, N H-EXP 5

Charge and Person-to-Person	**Credit Card, Collect & Person-to-Person calls** ...... Operator
Station-to-Station calls	Local numbers beginning with: 239, 242, 352, 363, 399, 446, 563, 585, 756, 827, 835, 847, 876 .... Number All other numbers .... Operator
Free calls	Directory Assistance .... Operator Toll Free 800 Numbers ...... Operator

AREA TYPE 2 EXP 5   Operator assisted rates apply to all toll calls from this telephone.

Instruction card set for a telephone that requires a coin deposit before any call can be dialed, and from which the operator dials all long distance calls.

No coin needed for Charge, SOS & Free calls.

1. 2. 20¢

Local calls Deposit 20¢ before dialing
Long Distance Dial all calls directly
  0+ needed for Charge & Person-to-Person calls
  1+ needed for Station-to-Station and Free calls

SOS dial 0 for Emergency help
SOS marque 0 para Emergencia

TSPS/DTF UPPER-EXP – 11

Charge and Person-to-Person calls	**Credit Card, Collect & Person-to-Person** Within this Area Code........**0+** Number Outside this Area Code.......**0+** Area Code + Number
Station-to-Station calls	Within this Area Code........**1+** Number Outside this Area Code.......**1+** Area Code + Number
Free calls	**Directory Assistance** Local........................411 Within this Area Code........**1+** 555-1212 Outside this Area Code.......**1+** Area Code + 555-1212 **Repair Service**...............611 **Toll Free 800 Numbers**......**1+** 800 + Number

TSPS/DTF STERLING LOWER-EXP – 11   Operator assisted rates apply to all toll calls from this telephone.

Instruction card set for a telephone from which credit card, collect, emergency, and other "free" calls can be made without depositing any coins, and from which long distance calls can be dialed by the caller.

Charge, SOS & Free calls only.   No coin calls.

Outgoing calls only

Charge calls	**Credit Card & Collect** Within this Area Code........**0+** Number Outside this Area Code.......**0+** Area Code + Number Operator assisted rates apply to all calls
Free calls	**Directory Assistance** Local (Queens)...............411 Within this Area Code........555-1212 Outside this Area Code.......Area Code + 555-1212 **Repair Service**...............526-9942 **Toll Free 800 Numbers**......800 + Number **SOS dial 911** for Emergency help

CAC EXP 14

Instruction card from a "Charge-a-Call" telephone.

The human factors staff at Bell, which includes 250 psychologists in various facilities across the nation, is involved in numerous other studies to simplify the process of communication. Since Bell psychologists think of customers as information users, they often ask novices to test computers. People who have little knowledge of such computer terms as *bytes* and *interface operations* are more concerned about whether the machine does what it is supposed to do than what makes it do the job. In short, novices want to be able to operate computers easily, without lengthy instructions.

What people want and how people function are also important aspects in the research and development of products at International Business Machines (IBM). The company considers itself "customer-oriented," and the human factors team is essential to understanding how people use IBM products, which include cash registers, computers, typewriters, and other business machines. As a company representative explained: "In an international company, knowing the user can often mean being aware of small differences. Cash register drawers that hold uniform American bills snugly won't do for many European countries where bills come in varying sizes. Some differences are more complicated. Japanese language keyboards may contain 100 times as many characters as European typewriters. Human factors people and development people worked closely on the [Japanese] keyboard to come up with a design that operators can easily learn."

Simplify. Simplify. Simplify. It is a goal stated over and over again in companies with human factors teams. And Eastman Kodak is no exception. The company has devoted a great deal of employee time and extensive resources to the research and development of easy-to-use cameras and copiers. Kodak disc cameras, for example, are considered the simplest ever made, and have been widely accepted by consumers. But years before the disc camera went on sale in 1982, the human factors staff, engineers, and marketing personnel were conducting tests and creating designs.

In 1976, the company began a basic study of how people use cameras and what kind of pictures they take. More than 30,000 prints by amateur photographers were carefully examined and the information was stored in a computer. "The analysis of this data told us what kinds of problems people were most likely to encounter and identified the picture-taking circumstances where they were most likely to occur," says Terrence Faulkner, a human factors engineer involved in the project from the beginning.

The information from the print evaluations helped the design team determine what the camera should do. At the same time, the researchers tested how the camera should look and feel. At first, designers thought the camera should have a shape something like a tiny pair of binoculars. A prism would reflect the image onto the disc film, which would lie horizontally. Three horizontal models and a number of vertical styles underwent hands-on testing. The vertical format proved easier to hold and was preferred by the test subjects. As a result, the vertical style became the design format for the new cameras.

Human factors personnel also wanted to minimize human error through the camera design. The print evaluation program allowed Kodak engineers to determine how often various human errors occurred. "We tackled the problem like detectives," says Faulkner. "For example, when I came across an underexposed print, I tried to pinpoint the specific reason for underexposure . . . ."

In a series of tests, the Human Factors staff began examining various shutter release mechanisms and flash positions as well as different styles of covers, handles, and camera shapes. In one test, volunteers (Kodak employees) "took pictures" with camera models made from styrene while the Human Factors staff noted whether fingers tended to stray over the flash or lens. In another test, camera movement was measured with a special device that detected any camera shake and automatically fed the data into a computer for analysis. Human Factors also explored what they called "holdability"—the general reaction to a camera design based on how it felt in the hands, whether it could be used with the left eye, while wearing eyeglasses, and a number of other factors.

One thing that emerged from the testing was the need for two styles of camera, one with a folding cover and another with a sliding lens cover. Another change was to make the lens opening rectangular instead of round, which helped improve contrast on prints. Kodak engineers also developed a scheme for a short recycling time between flash exposures so the camera would be ready for shooting again on almost a moment's notice.

When prototypes of the cameras were prepared, some 400 Kodak employees took disc cameras home on weekends for testing. Their findings, along with test results from hundreds more employees who volunteered for experiments, helped shape the design of the commercial disc camera, which has found its place in the Design Collection of the New York Museum of Modern Art.

(From the book *Ergonomic Design for People at Work,* Vol. 1, by Eastman Kodak Company. Copyright Eastman Kodak Company, 1983. Used by permission.)

60 / DESIGNING FOR CONSUMERS

Collaboration between human factors specialists and the other personnel involved with the development of a product can make a big difference in whether or not an object works well for people and has a good design. But technology also plays a major role in the design of products. As new technology becomes available, new materials and products develop. For example, plastic was available for wide use only after World War II, and the transistor became available commercially even later. From those two technological advancements have come countless products. In fact, it is hard to imagine everyday life without, say, transistor radios and the myriad of plastic utensils and coverings we use.

In the 1950s the first artificial satellite was launched. In the 1960s the space program in the United States went full speed ahead. Because of space technology, many consumer goods later were developed, including freeze dried foods, Teflon coatings, and insulated "moon boots" for winter wear. Recently, Harvard Medical School announced that it had developed an all-ceramic crown for teeth made from the same material used to manufacture heat shields for space capsules.

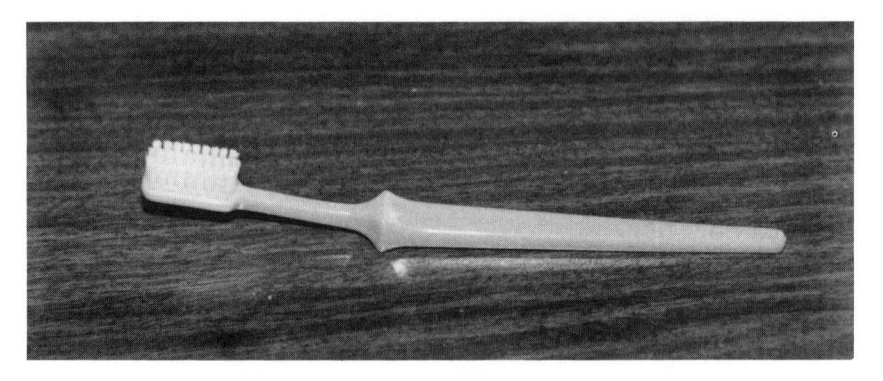

Ergonomists at a leading university helped develop the Reach toothbrush manufactured by Johnson & Johnson Company. The multigrip handle, angled like a dental instrument, and thumb flair were built in as part of the human factors design.

DESIGNING FOR CONSUMERS / 61

Video recorders and computers and other electronic equipment have become part of daily life in the 1980s. And the technology has spawned exotic robotics such as a "telepresence actuator"—an automated device under study by the National Aeronautics and Space Administration (NASA) that can manipulate and repair things in space while controlled by an operator on earth. With combined computer and video technology, engineers at Rochester (New York) Institute of Technology have created what is called a Landisc II. Using computer commands, pictures of thousands of pieces of real estate in Rochester are projected onto a video screen. This could be a forerunner to shopping by computer for property, groceries, or other products.

Designed for "friendly use," personal robots do a variety of tasks for humans. About 10,000-20,000 personal robots are now in use, and that total could jump to 1 million by 1990. This Hero I robot, manufactured by the Heath Company of Benton Harbor, Michigan, can pick up small objects with its arm. In the household of one of its creators/engineers, Kim McCavit, Hero I is programmed to rock the baby's cradle in a most soothing manner!

With new technology, new concepts are incorporated into new products, and some human factors specialists believe their discipline will be increasingly important. As Andrew LeCocq, a human factors engineer in Columbia, Maryland, has pointed out, the ultimate objective of the human factors profession is to deal effectively with "changes in the way people live, the way they do their jobs, and the things they must learn." LeCocq adds that these are "critical considerations for the future as systems and products become more sophisticated and users become more discriminatory."

Yet, no matter how sophisticated the system or what the technological advance, designing consumer products involves another important aspect: Safety. In 1972, Congress passed the Consumer Product Safety Act aimed "to protect the public against unreasonable risk of injury associated with consumer products." But there has been little reduction in personal injuries from some product lines. According to a number of studies, most personal injuries while using consumer products come about because of the need for more safety features in the design of those products.

"Many engineers who are responsible for design, testing, and quality control have not had the benefit of training in ergonomics and psychology. As a result, many products sold in the marketplace today reflect too high a risk acceptance for the ordinary consumer," says Joseph Ryan of National Standards Technology, Inc., a human factors consulting firm in Michigan. Ryan suggests that "human factors engineers can make a great contribution" in the safe use of consumer products by providing manufacturers with safety provisions that go beyond minimum standards set for certain manufactured items. Many times designers believe they have conformed to safety standards but the courts may find otherwise.

As Ryan points out: "The ultimate test of safe product design is whether the designer could have technically and economically

DESIGNING FOR CONSUMERS / 63

provided that degree of safety in the product to prevent injury when the product was used in a reasonable and foreseeable way." This means that a product may not always be used as a designer or engineer might have intended it to be used. And the "court recognizes that consumers will, and frequently do, use products in ways engineer designers never dreamed," Ryan says.

A product obviously designed with the user's safety in mind is this stepladder for handy homeowners, even left-handers.

64 / DESIGNING FOR CONSUMERS

So how can human factors specialists aid in the design of safe products? In the first place, ergonomists assume that human error is inevitable and misuse of products is highly probable. Usually, many injuries related to the use of consumer products can be anticipated and avoided if such factors are considered when the products are on the drawing boards. Perhaps a guard is not planned for a power mower because it is believed the user will always be careful when the blade is operating. But not allowing for a user's carelessness may make a manufacturer liable in the opinion of a court of law.

Ryan recommends that manufacturers "establish rigorous quality control procedures" to ensure safe products in the marketplace. Such quality control would include using safe and reliable materials, designing for possible misuse of products, testing adequately, and obtaining independent critiques of a product before it is released for sale. Ryan also stresses effective warning labels that alert consumers to possible product hazards.

Of course many injuries occur because consumers do not read labels or instructions. However, when a product is designed for an ordinary user, "it must be designed for a broad range of human beings with differing capabilities, intellects, and motivations," writes attorney John Messina in *Trial* magazine. He believes the design of the product has to be analyzed on the basis of what the ordinary consumer might do and whether that person would be exposed to risks by using the product.

HUMAN FACTORS IN TRANSPORTATION 6

Personal factors and their relationship to accidents have been under study for some time. As is well known, investigations have clearly shown that abuse of alcohol and other drugs causes a major portion of all traffic accidents. Fatigue, poor eyesight and hearing, emotional upset, and stress from the demands of driving are other factors that contribute to injury and death on the highways.

Yet, poorly designed roads, traffic controls, and the vehicles themselves are at fault in some serious accidents. In a widely publicized court case in New York, attorneys for 18-year-old Michael McHugh, a defendant in a fatal auto accident, tried to show that McHugh's car missed a curve on a rain-slick road because a warning signal was not working. But police who were at the scene claim McHugh was drunk and lost control of his car. It slammed into a concrete wall along a parkway in the Bronx. Four teenagers in the car with McHugh were killed.

McHugh was charged with four counts of manslaughter. During the trial, which took place nearly a year after the accident, "high tech" evidence was brought into the courtroom for McHugh's defense. Two physics professors, Arthur Damask and Arthur Paskin, from Queens College analyzed the physical evidence, putting weights, masses, speed, torque, and other data into a computer and creating a graphic simulation of what the data indicated happened at the accident.

The video simulation, which is somewhat like a cartoon film, included a small white box to represent McHugh's car and two curving lines for the roadway. During the 10-minute film, the graphics simulated the car missing the curve, sliding off the roadway, bouncing back, hitting an open utility manhole, spinning out of control, and crashing into the wall.

McHugh was acquitted of manslaughter charges. According to the jurors, polled after the trial, their vote for acquittal was based in part on the evidence presented in the video simulation. Professor Paskin says that since the McHugh trial he and his colleague have prepared similar computer simulations for five other automobile accident cases brought to trial.

Since so many factors in the operator-machine system can be blamed for motor-vehicle collisions, computer analyses may assist in finding the primary cause for a specific accident. "The traditional focus, especially on the part of law enforcement agencies, has been on the 'nut behind the wheel,'" writes attorney Messina. "After years of research, many human factors people now believe that the driver is really the most reliable component in the system. Numerous examples illustrate how a reliable driver, of average capability, can compensate for a badly designed vehicle or highway . . . ."

A driver can mount side mirrors on the car to improve visibility, or put reflective tape on the bumpers to make them more visible at night to other drivers. Careful drivers can compensate for a poorly marked highway by reducing speed and by braking sharply when going into unmarked curves.

Safety experts and human factors specialists have emphasized the need to improve driving environments—the roadways, the lighting, and signaling devices. New highway signs are needed, too, but drivers can be overwhelmed at times if there are too many sources of information at once.

HUMAN FACTORS IN TRANSPORTATION / 67

One way to prevent driver overload and confusion is the use of standard traffic signs. The U.S. Department of Transportation (DOT) requires states applying for federal highway funds to adopt the national coding system for traffic control devices. The yellow diamond, for example, is the uniform color and shape for warning signs that caution drivers about hazardous road conditions, such as slippery pavement ahead. Directions and information about routes are to appear on green, rectangular signs. The eight-sided red marker is already a well-established symbol—the standard **STOP** sign.

Cattle Xing	Farm Machinery	Divided Highway Ends	Two-Way Road to Divided Highway.	Two-Way Traffic Ahead.
Overpass Ahead	Merging Traffic.	Traffic Signal Ahead.	Slippery When Wet	Advance Warning.
Pedestrian Crossing.	Long or Steep	Lane Ends.	Advance Warning.	Bike Xing
Right Turn Ahead.	Winding Road.	Reverse Turn Ahead.	Reverse Curve Ahead.	Curve Ahead.
Deer Crossing Sign.	Crossroad.	Side Road.	"T" Intersection.	"Y" Intersection.

The diamond is the standard shape for warning signs in many states.

68 / HUMAN FACTORS IN TRANSPORTATION

The sizes of letters and numbers on highway signs and where the signs are located can affect the driving environment, studies have shown. Research also indicates that divided highways (such as the U.S. interstate highway system) and lighted roadways at night significantly reduce accidents.

When interstate highways and local street exits are clearly marked on overhead signs, traffic flows more safely through a large city like Chicago.

The design of safe and comfortable vehicles has an impact on accident prevention as well. In recent years, some automobile manufacturers have hired ergonomists to provide data on human characteristics and measurements that should be considered when designing car interiors. At General Motors' Fisher Body Division in Warren, Michigan, a staff of twenty-five, headed by Susanne Gatchell, Ph.D., study and test the way people use or operate controls, locks and handles, seat adjustments, and many other features. Called the Human Accommodation Group, the ergonomists at GM say they "sweat the details"—they test, and test again, to compile data on such factors as the rate at which people make mistakes using new controls, the comfort of the driver, and the ease with which seat belts can be fastened and the tension adjusted.

For the tests, design engineers use a mock-up—a simulated car interior—and call on volunteers within GM for experiments. Each year 300-350 employees take part in testing. Many more GM people volunteer for anthropometric measurements. Using a sophisticated instrument called an anthropometer, engineers measure arm length, shoulder width, seated height, hip width, and other body dimensions. The data from 2000 or more subjects then is used to create a life-size drawing of a car interior that will fit 95 percent of the population.

Because average foot size has increased for the American population, designers at GM also use an additional human factors aid—a model of a shoe which engineers have dubbed "bigfoot." The model helps designers determine the size of pedals and the space needed under the instrument panel.

As in office furniture manufacturing, seating is an important factor in car design. The comfort of drivers may depend on how well their car seats fit. Ergonomists at GM continually test seat designs and recently found seat tracks needed to be lengthened so that a seat could be moved forward or back to fit someone as short as 4 feet 11 inches (the 5th percentile female) or as tall as 6 feet 1 inch (the 95th percentile male). Some car designs with such features as power seats that move up and down as well as back and forth accommodate even shorter or taller folks. Power seats can also be tilted to provide more headroom.

Along with comfort factors, GM ergonomists (as well as those in other auto manufacturing plants) study and recommend safety features for cars. One of the new safety devices for GM models is an extra brake light placed above the back seat. The light is designed to warn other drivers following several cars back that the driver ahead is braking. A plastic cover for the front windshield is another safety feature being developed. If someone hits the windshield in an accident, the cover would be a protection from flying or shattered glass.

Designing for safety has long been a matter of concern for ergonomists, but the subject also has created much public controversy. In a 1966 ergonomic study titled *Man and Motor Cars*, British author Stephen Black pointed out that if people inside a car are going to be protected, the only purpose for the design of an auto body interior should be safety. But he asks, "Will safety sell?"

It is a question many auto manufacturers have posed, since car buyers seem to prefer styling over safety features. Consumers are often "sold" by status appeal—a car is advertised as a way to project a certain image. Just a look at TV commercials and magazine ads for cars will make it clear that different cars are designed to project images of success, sexiness, power, sportiness, and other characteristics. Few ads emphasize a car body that will absorb great impact, or a superior braking system, or controls that are easy to distinguish from each other. Shape, color, contour, convenience, gas mileage, "go-power," and other economic or cosmetic features get more attention than lifesaving factors.

Even such safety devices as seat belts and inflatable restraints (air bags) have been difficult to "sell" to the general public, in

HUMAN FACTORS IN TRANSPORTATION / 71

spite of the fact that many studies show the devices save lives and prevent serious injuries. Since the 1960s, when air bags were first introduced, these protective cushions, designed to inflate on impact, have been the subject of countless debates and much confusion.

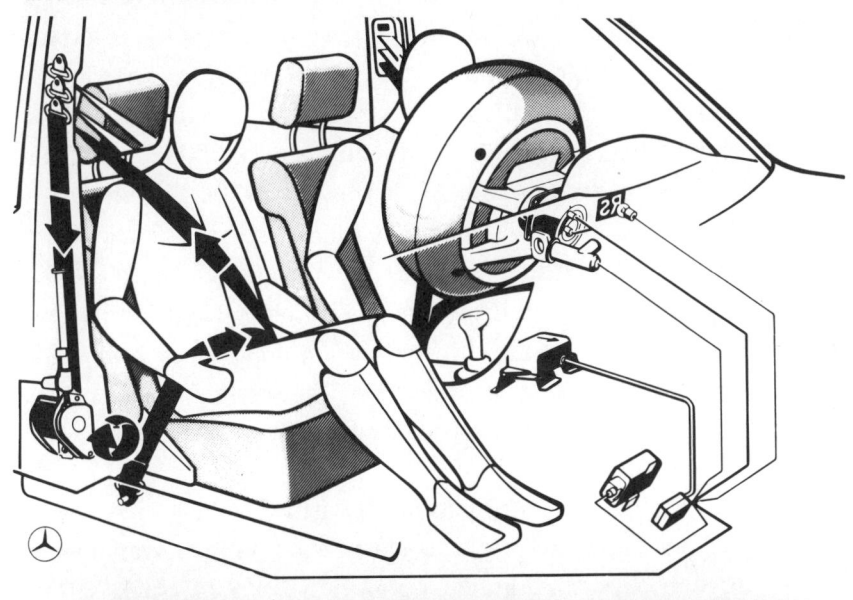

The Mercedes-Benz Supplemental Restraint System works with the seat belts to provide additional protection in a major frontal crash. Known as SRS, it includes an air bag (housed in the steering wheel) and a knee bolster for the driver, and an emergency tensioning seat belt retractor for the front passenger.

Auto manufacturers have been opposed to air bags, insisting that they are too costly to install and would raise the price of cars, which would not be acceptable to buyers. Other critics of air bags argue that the cushions would inflate when not needed, such as when slamming the hood of the car or getting bumped in a parking lot. Some fear that the gas that inflates the protective cushions would be toxic and believe that more testing is needed.

Those who favor inflatable restraints point to the fact that General Motors, Ford, and Volvo equipped 12,000 cars with air bags for demonstration purposes in the mid-1970s. By July 1983, those cars had travelled nearly a billion miles and had been involved in 267 crashes which were severe enough to inflate the bags. According to the Insurance Institute for Highway Safety, "The reductions in life-threatening injuries, compared with unrestrained occupants in similar cars, in severe frontal crashes have been very impressive, continuing to run within a few percentage points of sixty-five percent."

The Insurance Information Institute, which has been campaigning for many years to have air bags made available to car buyers, says that public opinion polls show such safety devices would be popular. The organization also points out that laboratory and field tests have revealed no toxic effects or other hazards from air bags. In addition, costs have been analyzed, and it has been found that the price for air bags could range from $100 to $200 apiece if produced in quantities of one million or more.

The Department of Transportation (DOT) has long advocated that air bags or seat belts be installed in all cars and finally passed such a ruling in July 1984. But the law is complex. It provides that by 1989 all new cars made in the United States would have to be equipped with air bags, seat belts that close automatically, or other "passive restraints," which could be some type of protective device not yet developed. However, according to the DOT ruling, if enough states, representing two-thirds of the total

HUMAN FACTORS IN TRANSPORTATION / 73

U.S. population, require that seat belts be worn, the federal air bag requirement would be void. New York state passed a mandatory seat belt law early in 1984. New Jersey, Illinois, and Michigan followed suit soon afterward, and many other states are considering similar legislation.

Some consumer groups and insurance companies believe the DOT ruling does not really protect the public. A third of the population would not be covered by seat belt laws and air bags probably would not be installed in many cars.

Since the automobile is the favorite mode of travel in the United States, other forms of transportation do not always receive as much attention in terms of people-oriented designs. Yet, public transportation in many parts of the nation could benefit greatly from ergonomic input. Millions of people have to commute by train, bus, or subway to city centers and must

---

An "idea car" is this Astro III by Chevrolet. The jet-styled, two-passenger experimental vehicle is designed to probe future possibilities for turbine-powered personal automotive travel. Designers see the Astro III as a high performance vehicle suited for travel on restricted access or possibly systems-controlled interstate highways of the future. The rear-mounted gas turbine engine and the positioning of passengers close to the ground contribute to a low center of gravity essential for good handling and maximum stability with the tricycle-type wheel arrangement. Chevrolet has no production plans for the car, but is carefully evaluating the design concepts.

adapt to the transportation system rather than having a system designed to accommodate riders. To many commuters, better safety and security systems, less crowding, more comfort, cleaner and more convenient station facilities, and on-time schedules would make public transportation more suited to users.

Commercial air travel, although not as convenient as some would like, is considered thirty times safer than highway auto travel. But critics of air traffic safety argue that much can be done to make the skies even safer. Some worry that rising costs of airline operations could mean cuts in safety programs and the number of engineers and technical staff who guard against aircraft hazards. Better aircraft design and studies of air traffic control are part of human factors research on air travel safety.

Measures have been proposed for designing safer airplanes, including features that would protect passengers from disastrous fires—such as fire retardant materials for seat cushions and carpeting, and fire extinguishers in cabins. According to the National Transportation Safety Board (NTSB), of the total number of passengers killed in plane crashes, one out of five dies from fire, smoke, and toxic fumes.

A number of fires in aircraft aloft have been caused by people who have dropped lighted cigarettes in restroom wastebaskets, even though the Federal Aviation Administration (FAA) has banned smoking in aircraft lavatories since 1973. The NTSB has long recommended that smoke detectors be installed on commercial airliners. Yet, to date the FAA has not made any such ruling. However, several airlines, including Pan Am and Air Canada, have already installed smoke detectors in aircraft lavatories.

Another major concern is pilot performance. Studies are underway by NTSB to determine what human factors may contribute to "pilot error." Flying across time zones, for example, can affect eating and sleeping habits and could have effect on

HUMAN FACTORS IN TRANSPORTATION / 75

the pilot's ability to remain alert. In some new jets, pilots must monitor highly sophisticated computer systems with little "hands-on" flying required. There are some concerns that "flight managers" (as pilots of these computerized aircraft are called) will become apathetic and will not respond well in emergencies.

Whatever the means of transportation, numerous human factors have to be considered if people are going to get from

An engineer at NASA's Langley Research Center (Hampton, VA) is shown using a ride quality meter. He is conducting a test that simulates vehicular motion. Sensors report information on noise and vibration levels. The results are then processed by a computer and printed out. Designers can use the computer data to improve passenger comfort in air, sea, road, and track vehicles.

place to place comfortably and safely. Pedestrians and bicyclists need pathways protected from motor traffic. Motorcycle riders need protective clothing and headgear. Passengers and crews on ocean liners expect to be guided into and out of ports without collisions. Vacationers in recreational vehicles hope to be as safe in their rolling homes as in their permanent residences. On land, sea, and in the air, ergonomic data can help improve the quality of travel and the safety of travelers, which so often depend on the design of machines, devices, and total transportation systems.

DESIGNS FROM SPACE RESEARCH 7

In no field has the use of ergonomic principles been more important than in the aerospace industry. For a space vehicle, everything must be provided, including food and methods for disposing of human waste, to make the closed system habitable for people. Researchers also have had to determine how zero gravity affects people in space and how to use controls, equipment, and tools in a weightless state.

Authors Eloise Engle and Arnold Lott, who wrote *Man in Flight: Biomedical Achievements in Aerospace*, describe in their book many of the human factors that had to be considered through the history of space research. One of the first design problems was to develop a pressure suit that would allow pilots to fly at high altitudes where low-pressure conditions could cause death. According to Engle and Lott, "The first serious effort to design and build a workable pressure suit was instigated in 1933 by Mark Ridge, an American balloonist. He prevailed upon physiologist John Scott Haldane (who had suggested such a suit in 1920) and Sir Robert H. Davis, a British diving suit specialist, to design and build him a protective suit."

Ridge successfully tested the suit a number of times in a low-pressure chamber that simulated high altitudes, but he was unable to raise the funds to buy a balloon or aircraft to try out the pressure suit in a real-life experiment. All through the 1930s, however, as aircraft were developed to fly higher and at ever-greater speeds, a number of countries, including the United States and the Soviet Union, designed and produced high-altitude pressure suits. But the problem was making a pressure suit that was "wearable and workable" at the same time.

In the mid-1930s, the Soviets developed a protective suit resembling a deep-sea diver outfit, and by the 1950s, Americans were testing a full-pressure suit that eventually became the space suit worn by Project Mercury astronauts in the 1960s. Engle and Lott describe the first space suit as "an emergency back-up to the . . . environmental control system" of the space capsule itself.

Along with advances in space suit design over the next two decades, the environment of the space capsule was refined and made more habitable. For Gemini flights in 1965 and 1966, astronauts were able to wear lighter-weight suits and could easily remove their fabric hoods, and boots and gloves. When the pressurization in the cabin became more reliable, astronauts could use regular flying suits while in orbit. During Apollo missions to the moon, astronauts wore their space suits only for launching, during the moon orbit, while exploring the lunar surface, and reentry into the earth's atmosphere.

Ergonomists and biomedical researchers have contributed much to advances in space technology since the 1950s and 1960s. Some of those advances may seem minor, as in the case of a toothpaste developed for outer space. Most people would hardly give a thought to how astronauts brush their teeth. But handling that part of daily hygiene—a simple task on earth—caused problems in space.

For one thing, during the early space flights, most of the spacecraft was packed with equipment needed on the mission. The area provided for astronauts was small and cramped. No one

DESIGNS FROM SPACE RESEARCH / 79

even considered including a basin or any other container that might be used for the toothpaste that is spit out after brushing. Besides that, in zero gravity, the astronauts would be unable to get the foam and liquid into a basin!

Long before extensive space flights were underway, the National Aeronautics and Space Administration (NASA) found a researcher who was able to develop a foamless toothpaste that could be swallowed. It can be used without water, has a pleasant taste, and causes no harm when ingested. Astronauts and their families have been using the toothpaste for many years, and it is now available as a commercial product called NASAdent, produced by Scherer Laboratories, Inc. of Dallas, Texas.

80 / DESIGNS FROM SPACE RESEARCH

An ingestible toothpaste has additional benefits. It is an important aid for some hospital and nursing home patients, paraplegics, and others who may need help brushing their teeth. The product is also useful for very young children who are just learning how to brush their teeth and often swallow toothpaste because it "tastes good." Now with an ingestible toothpaste kids can safely eat it!

Of course, many more complex space problems have been tackled by ergonomists and other researchers. In planning for the Space Shuttle program, for example, one of the design problems involved development of insulation materials that would minimize fire hazards. Long before the first shuttle launch, researchers were experimenting with polymide foam materials that resist fire. Today these materials are used not only in the construction of spacecraft but also in building aircraft, ships, and vehicles such as buses, trains, and rapid transit cars.

The first flight of the Spacelab orbital laboratory was launched on November 28, 1983. The lab was part of the ninth flight of the Space Shuttle. A crew of six was aboard.

In the design of the Shuttle Orbiter, human factors specialists also had to make provisions for the craft to carry not only pilots as in the earlier space flights but also passengers. People of both sexes and different ages would be aboard. The crew would move from pressurized areas to nonpressurized areas, and designers had to determine what effects such movements would have on astronauts. A variety of tasks different from those performed on other spacecraft would have to be accomplished. These and many other factors had to be taken into account in the design of clothing, equipment, work and living areas, and life-support systems on the Orbiter.

A cotton two-piece garment that can be worn most of the time when the shuttle is in orbit.

82 / DESIGNS FROM SPACE RESEARCH

Life-support systems also had to be developed for astronauts who ventured outside the spacecraft and maneuvered in space without tethers or umbilical lines connected to the spacecraft. A spacesuit/backpack called an Extravehicular Mobility Unit (EMU) supplies astronauts with oxygen. It also removes carbon dioxide, controls the temperature, and protects against meteoroids. Attached to the EMU backpack is a Manned Maneuvering Unit (MMU), which is an orbital propulsion system with nitrogen gas jets that provide the thrusting impulses. The astronaut uses hand controls built into the unit's armrests to maneuver in space. Mission Specialist Bruce McCandless was the first astronaut to use the MMU during the tenth Space Shuttle mission in February 1984. But the EMU and MMU spacesuit/backpack systems have been used in later missions, and NASA expects they will be used in many future missions to retrieve, repair, and service orbiting satellites.

A maneuvering unit designed to give astronauts a complete and safe environment for work outside a spacecraft.

In recent years, as part of the Space Shuttle Program, NASA has developed the Spacelab, an orbital laboratory. The Spacelab provides opportunities for a variety of experiments in space, some of which are conducted by scientists and engineers who accompany the astronauts. During the first Spacelab flight in November 1983, dozens of separate research projects were conducted. These included experiments to learn what effects microgravity and radiation have on plant systems, what causes people to have motion sickness in space, and what can be done to help people adapt to weightlessness.

In several recent Space Shuttle flights, many experiments have involved "orbital materials processing technology." Using this technology, scientists are exploring ways to develop new treatments and possible cures for many diseases. The scientists use substances produced by the human body—cells, enzymes, hormones and proteins. Earth's gravity allows only a very small sample of these substances to be extracted from biological materials. But in gravity-free space, large quantities of biological materials can be separated. The space environment also offers high levels of purity. Thus Ortho Pharmaceutical Corporation, a division of Johnson & Johnson, working in conjunction with NASA and McDonnell Douglas Astronautics Company, plans to operate a computer-controlled pharmaceutical processing plant in space. The fully automated "space factory" will orbit the earth and could be in operation by the end of the 1980s. NASA says the facility "will be serviced by Shuttle crews who would deliver raw materials and collect separated products."

Such orbiting factories will be part of a future space station planned for the mid-1990s. The space station will be a facility not only for materials processing and development of pure pharmaceuticals. It will also be a laboratory for research in earth and life sciences, and in such fields as astrophysics and communications. Exploration of the solar system will also take place from space stations.

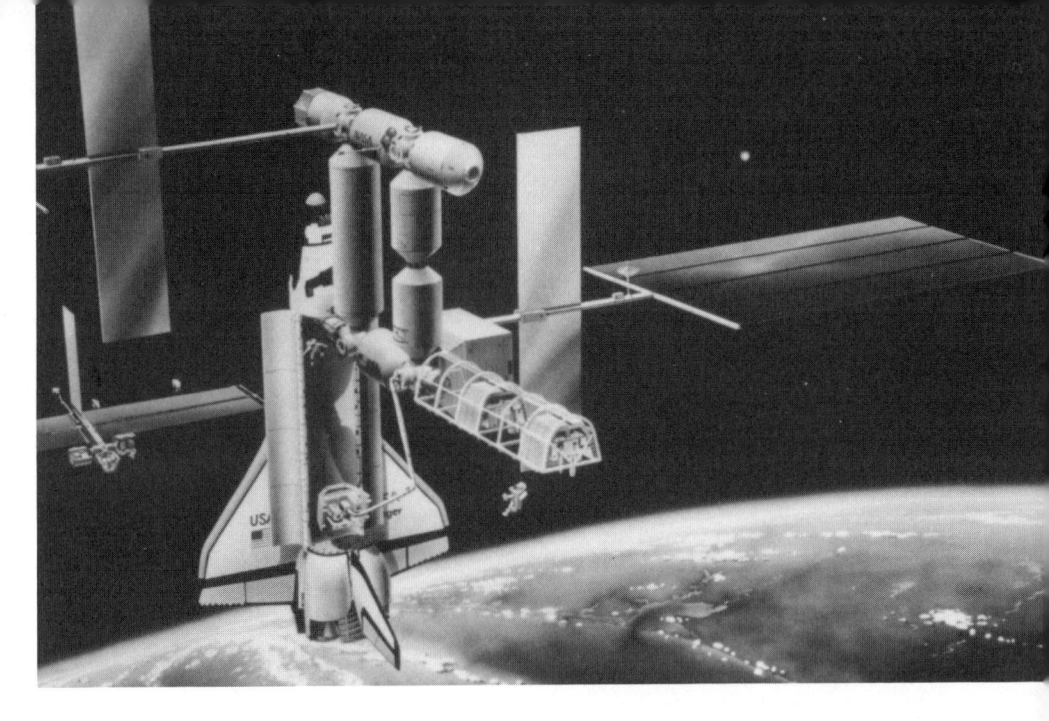

The exact design for U.S. space stations has not been determined yet, but these illustrations show how two types of space stations might look. The one above has an open rack-like structure for experiments that require exposure to the space environment. Shown below is a model of a more advanced space station. It could be expanded with add-on modules or sections. The station would also include hangars (center foreground) for Orbital Maneuvering Vehicles, planned "space tugs" that could move satellites and other objects in space.

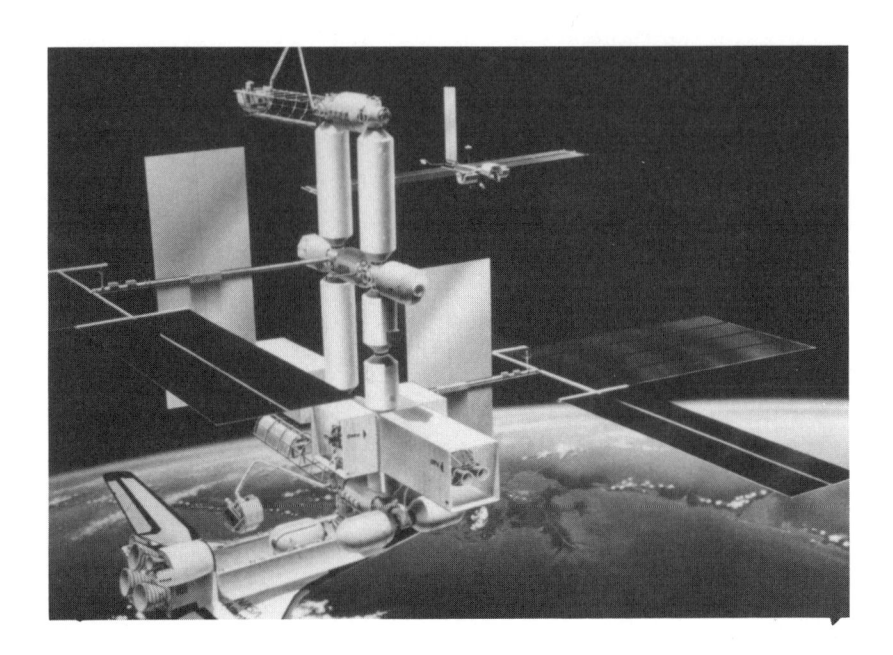

DESIGNS FROM SPACE RESEARCH / 85

To tackle the myriad design problems that must be solved in order to put people and experimental stations in space, human factors specialists have had to compile a vast amount of anthropometric information—data on the sizes, shapes, and motion characteristics of many different types of people. The information is available through a data base at the Johnson Space Center in Houston. However, officials at the center believed the nation would benefit if there was wider distribution of the material. Eventually, NASA made arrangements to publish the data on human characteristics in a three-volume work called *Anthropometric Source Book*. It includes tabulated characteristics of sixty-one different population groups in the United States, Europe, and Asia. The reference work also covers such subjects as variations in body sizes, arm and leg reach, strength, and joint motion. There are guidelines for clothing and workplace design as well as information on how to analyze anthropometric data. Since the data apply in many nonaerospace fields, the source book has been used in a variety of professions and industries.

Kodak's human factors specialists have used the data to design comfortable and productive workplaces such as the one shown below which is used by employees who process disc film negatives.

Other companies have used the anthropometric data to design protective clothing, manufacture a variety of safety and comfort features for aircraft and land vehicles, and develop equipment to aid the handicapped.

One special device for the severely handicapped has come about because of the Lunar Land Rover. Apollo astronauts used the Lunar Rover to explore the surface of the moon in the early 1970s. The astronauts operated the vehicle with a "joystick" since the Lunar Rover had no gas and brake pedals or steering wheel. All operations could be accomplished with one hand. The technology which helped produce the Lunar Rover has enabled Johnson Engineering Corporation of Boulder, Colorado, to manufacture a vehicle control system called UNISTIK. The UNISTIK system is designed to "operate all primary controls necessary for driving and controlling a highway vehicle," NASA says, thus a vehicle "can be operated by handicapped persons who have no lower limb control and only limited use of upper extremities." Commercial production of the system began in late 1985.

Other spinoffs from biomedical and human factors research for space missions have provided numerous safety dividends and protective devices for everyday use. For one, a personal security system has been developed based on space telemetry technology, in which electronic signals are transmitted by radio to a receiver that decodes the signals. The heart of the security system is an ultrasonic transmitter the size of a pen.

Prison guards can use the device if an emergency arises in a cell block. The transmitter sends a radio signal to a receiver (usually mounted in a ceiling) that is connected to a central display panel in the prison. Within other buildings such as offices, schools, apartments, and retirement centers, the device can be used by occupants to send an alarm or call for help. It may also be used to activate lights or machinery, open a door, or initiate an automatic telephone call.

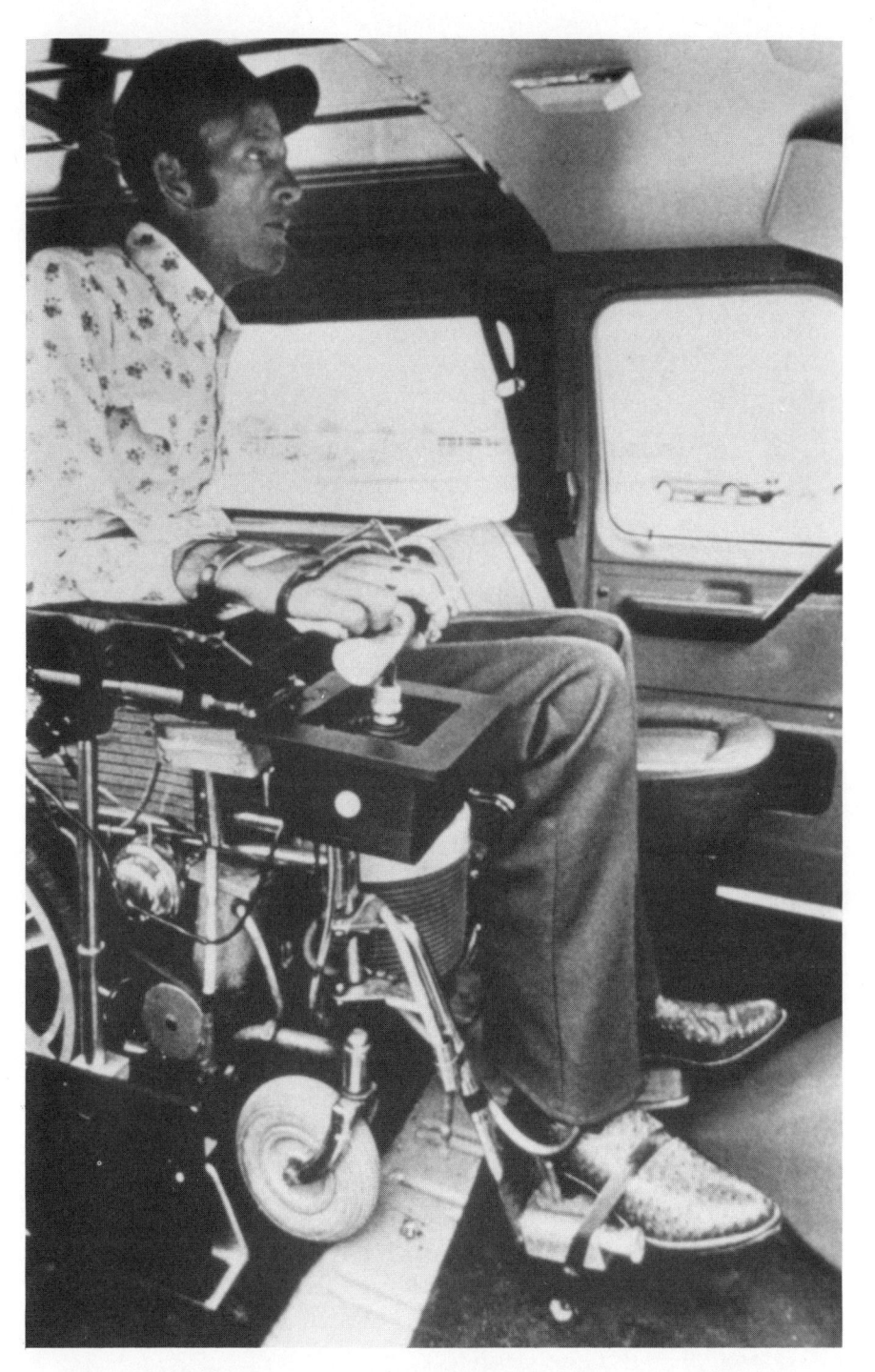

Johnson Engineering's UNISTIK vehicle control system is a spinoff of the system Apollo astronauts used to control the Lunar Rover. UNISTIK uses a joystick instead of a steering wheel, brake pedal, and throttle pedal.

88 / DESIGNS FROM SPACE RESEARCH

NASA research led to the redesign of safety equipment for fire fighters. The breathing apparatus used to protect fire fighters from smoke inhalation has been heavy and difficult to carry. But a new air bottle, 30 to 40 percent lighter than the old device, has been developed from technology for rocket motor casings. NASA engineers also redesigned the pack frame and harness so that the weight was better distributed, making breathing equipment seem even lighter.

A line of fire-retardant coatings for buildings is being produced because of a spinoff from spacecraft construction. The coatings were previously used as part of the heat-shielding system for spacecraft. Fireproof insulation and fire-retardant materials for clothing, furniture coverings, carpets, drapes, and curtains are other developments from aerospace research.

NASA engineers developed another safety device, the tornado detector, which is light-sensitive and attaches to a TV set, tuned to any unused channel. By using the electronics of the TV set, the system detects any tornado within 18 miles and sounds an alarm. It cuts off automatically when the storm moves out of range.

Hundreds more innovations from space research have improved the safety, health, and comfort of countless Americans and people in other parts of the world. Satellites in space help weather forecasters; lives and property have been saved because of adequate warnings of hurricanes, tornadoes, and tropical storms. A variety of devices developed from space technology help protect the environment. Numerous medical spinoffs from aerospace research have saved lives and have improved health care for millions.

How does information about space technology get to people who will use it to develop new products or processes? In 1962, NASA established the Technology Utilization Program with headquarters in Washington, D.C., and field centers throughout the United States. Specialists in the centers find new ways to apply space technology, then they encourage private businesses and government facilities to make use of the information.

A key factor in obtaining and transferring information is tapping into NASA's computerized storehouse of more than 10 million documents on technical subjects. NASA also provides businesses with many types of computer programs that were originally developed to analyze aeronautical designs, data from satellites, and many other operations. With little or no modification, a variety of companies can use the programs at a much lower cost than developing new computer software.

In addition, NASA engineers and technicians work directly with companies applying new space technology. They might help set up a human factors laboratory to study how vibration and noise can affect the performance of someone operating heavy construction equipment. Or specialists might help a company produce new designs for airplanes, hair-styling appliances, sunglass lenses, sewage treatment plants, or any number of other goods and services. In short, NASA's space technology is being recycled, transferred, or applied to make products, places and processes fit for general human use on earth.

PEOPLE SPACE ON EARTH 8

- The new corporate headquarters in Southern California sat atop a hill overlooking the Pacific Ocean. But workers complained because there were no windows in the building for enjoying the view. Some of the office staff also grumbled about the "cold, impersonal" look of the interior, and others blamed the air conditioning system, which had to operate all year around, for headaches and other ailments.

- The old school was completely remodeled and the playground enlarged. But after the refurbishing, teachers, parents, and some students were upset because of the vast expanse of blacktop used for the playground surface and the stark, steel play equipment that had been installed. Many wondered why grass and sand were not included in some play areas. And people often asked why the jungle gyms and slides were not the colorful new designs used on playgrounds for other new schools.

- The street had been a busy place with people congregating in or near small stores and shops, or meeting friends at the corner drugstore or neighborhood cafe. But one by one the buildings in the old neighborhood were sold and eventually torn down. A new supermarket was constructed and most customers liked the vast array of products to choose from. But many of the longtime residents in the neighborhood were disappointed because they no longer had "gathering places." The supermarket and its exterior environment were designed to move people in and out quickly, not to encourage social exchange. Few people inside or outside of the market stopped to chat.

These are just a few examples of the way designs for "built space"—buildings and constructed environments outdoors (such as parks)—fail to meet the needs of the people who use them. Until recent years, architects, interior and landscape designers, city planners, and others involved with construction did not necessarily plan with users foremost in mind. Instead, the main thrust was to create a beautiful artistic setting or a building meant to be looked at and admired as a work of art. The functional or practical aspects were a secondary consideration.

Ergonomics input has found a place in the design of some office buildings and a few industrial workplaces, as previous chapters have noted. But behavioral scientists (psychologists, sociologists, and others), designers, and architects are beginning to work together to "humanize" not only buildings, but the total life space, which can range from a single apartment to a huge downtown development project.

Robert Sommer, Ph.D., a psychology professor and the director of the Center for Consumer Research at the University of California at Davis, has been a consultant for architectural firms for over twenty-five years and calls the cooperative efforts between design and behavioral science social design. In a 1983 publication with that title, Sommer described social design as "working with people rather than for them; involving people in the planning and management of the spaces around them; educating them to use the environment wisely and creatively to achieve a harmonious balance between the social, physical, and natural environment; to develop an awareness of beauty, a sense of responsibility, to the earth's environment and to other living creatures . . . ."

As Sommer points out, social design became part of the human rights movement of the 1960s when social scientists and designers began working together on revitalizing inner-city neighborhoods and providing housing or other facilities for the handicapped, the elderly, and the poor. But social design also includes a concern for improved office buildings, civic centers, public parks, schools, and even such facilities as fire stations and sewage treatment plants. In short, "the goal of social design is to produce buildings and neighborhoods that suit the occupants," Sommer writes.

But studies and data collected on human needs and behavior are not enough. Although social designers use information from interviews and survey questionnaires, their main emphasis is on making sure that the people who will use built space are allowed to participate in the planning process. Sometimes it is not possible for users of occupied space to be present at planning sessions, or different groups may not be able to represent themselves adequately, especially if there are language barriers or handicaps. So social designers act as advocates for users or occupants of built environments.

94 / PEOPLE SPACE ON EARTH

The advocate role is also a part played by many human factors specialists. In fact, the ergonomist and the social designer have similar approaches to improving the fit between people and their environment. But human factors designers may concentrate on improving human productivity and reliability along with making users of occupied space more satisfied with their surroundings. Overall, though, the aim is the same: creating people-oriented space.

Yet, with so many differences in the way people work, play, and live, how can social designers or human factors designers apply their expertise and achieve their aim? What are some of the "nuts-and-bolts" aspects in the design of built space?

For one, human factors designers can help occupants of different types of buildings determine what they need or want in room size and arrangement. In an apartment building for the

---

"Kidspace" in a public library could include this type of "Toddler Area" designed to accommodate very young children while parents browse in the book stacks.

elderly, for example, builders might believe that residents need one large community room, but surveys could show that several rooms are needed for social activities. Perhaps designers assume that older people are inactive, when in fact the older residents may want such facilities as outdoor walkways and bicycle paths.

Lighting can be a major consideration in many buildings, particularly in workplaces where people perform exacting visual tasks such as assembling small manufactured parts or in hospitals where doctors perform delicate surgical procedures. Ergonomists also assist in the design of systems to control the atmosphere in built environments (which range from underground homes to space capsules) by providing information on how people are affected by loss of gravity and by temperature, humidity, and other atmospheric conditions.

Noise control is another factor that has to be considered in such buildings as schools, offices, and factories. Only in the past

A high-density open-plan work environment provides every employee with ample work and storage space while at the same time maintaining much-needed privacy.

20 to 25 years has noise been regarded as an environmental pollutant—not visible, but a potential hazard to hearing, and thus to health nonetheless. The primary focus has been on control of noise produced by machinery in industry and on how to protect workers from hearing loss due to high noise levels.

Usually, noise is measured with sound-pressure meters that record sound in decibels (dB). The federal regulatory agencies, Occupational Safety and Health Administration (OSHA) and Environmental Protection Agency (EPA) set standards for safe noise levels in industry which limit workers to 90 decibels (dB) during an eight-hour workday. For shorter work periods up to 100 dB exposure is allowed. However, some ergonomists say that even noise levels above 80-85 dB are dangerous. Workers are often required to wear protective devices such as ear plugs and muffs to shield them from damaging noise levels.

Although the danger of hearing loss from noise is not a threat in schools, high noise levels do cause other problems. Some studies show that disruptive noise—sounds from the gym or music room, from construction outside, from jets overhead, and so forth—have an effect on learning and teacher stress.

In one survey, teachers said they believed noise levels in everyday life have increased to the point where many students are uncomfortable with quiet. Silence seems to irritate them, and students beg for some kind of noise so they can concentrate. Still, adapting to noise and accepting noise as a "friendly companion" do little to enhance the physical and mental health of young people. It might be a good role for ergonomists to make the case for "peace and quiet": to educate people on the real hazards—ranging from hearing loss to heart failure—of long exposure to high noise levels.

Stress in the office caused by clatter from computer printers, typewriters, and other machinery has received much attention. However, as mentioned in Chapter 3, some electronic offices are being "humanized" with a variety of sound-proofing measures from carpeted floors to special covers to mute sounds from printers.

Stair safety is another human factors problem in many buildings. Each year in the United States, accidents on stairs cause about 800,000 injuries that require hospital emergency treatment, and between 2 and 3 million people may sustain minor, unreported injuries from stair accidents. Although children and young people have the highest rate of accidents on stairways, the elderly often suffer permanent disabilities from falls on stairs, and 85 percent of the 3000 to 4000 deaths annually due to stair accidents occur in the over-65 group.

According to Jake L. Pauls of the National Research Council of Canada, "stair safety research, with special attention to human factors, has been conducted in Britain, Canada, Japan, Sweden, and the United States." Design recommendations, resulting from the research, have helped bring about changes in building codes and have helped set standards for stair safety. "In general," Pauls says, "research has shown that . . . better stair design and construction [can] reduce human error on stairs and make stairs more 'forgiving' when an error occurs thus reducing the number and severity of accidents." Pauls cites three basic rules for stair design that would help eliminate hazards: 1) Make stairs large enough so that footing is secure; 2) Provide handrails that can be grasped easily; 3) Make sure the stairs can be seen.

Human factors principles also can be applied to outdoor space. Specialists may learn that the reason a play area for small children is not being used is that it is located in an area that parents believe is unsafe or is not easy to get to from nearby apartments. Or they could learn that a plaza is uninviting because too little provision has been made for seating. As an example of good planning, they might learn that people in a neighborhood need a park more for strolling and "looking" than for activities such as picnics and ball playing, and build the park for that purpose.

Open space and built space with kids in mind take into account the needs of people living in apartment complexes.

It is impossible here to describe all the ways that open space could be designed to suit users. But fortunately more and more city planners, landscape architects, and builders are aware that open spaces which fit people improve the quality of urban life. Whether outdoors or inside, built spaces should be "people places"—functional, comfortable, usable as well as pleasant, and enjoyable to behold.

HANDIABLE DESIGNS 9

People with the most pressing need for livable space and usable products are those with physical and mental limitations. In fact, as Tom Cannon, a human factors consultant in Colorado who has designed equipment for the visually impaired, put it, "No segment of the population suffers more from neglect of human factors requirements in product design than the severely handicapped."

Nearly 3 million Americans are extremely limited in their vision, hearing, motor activity, or mental abilities. And the Office of Technology Assessment (OTA), which analyzes and advises Congress on technological issues, says that anywhere from 15 to 85 million people are in some way limited in their ability to perform one or more important life functions such as eating, speaking, or walking. The disability can be present at birth or be caused by a disease, injury, or aging. Not all disabilities create severe handicaps, however. A disabled person may become handiable (able to function independently) with the use of various types of devices, many of which have been developed as a result of new technology.

Yet, even though technology exists to help the handicapped, there are still many problems in pulling together human factors information on the handicapped in a form that can be used by designers. Why the lack of data? Cannon says that organizations

100 / HANDIABLE DESIGNS

set up to aid people with handicaps are "oriented toward attacking the source of the disability rather than measuring the resulting functional impairment." And even when information on the abilities of the handicapped is compiled, it may not be widely distributed. Cannon believes that if industrial designers were more aware of the needs of the handicapped, products would be designed to accommodate their more limited abilities. Such products would help not only the impaired—improved designs would also benefit the nonhandicapped population, he says.

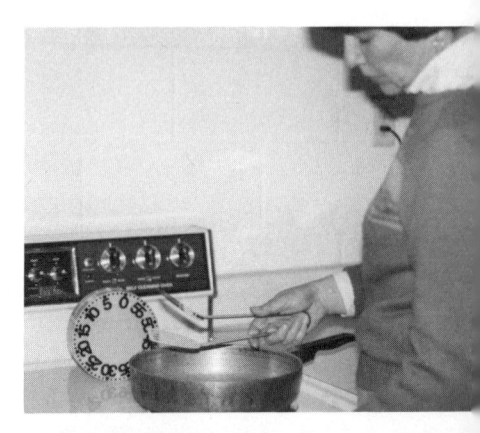

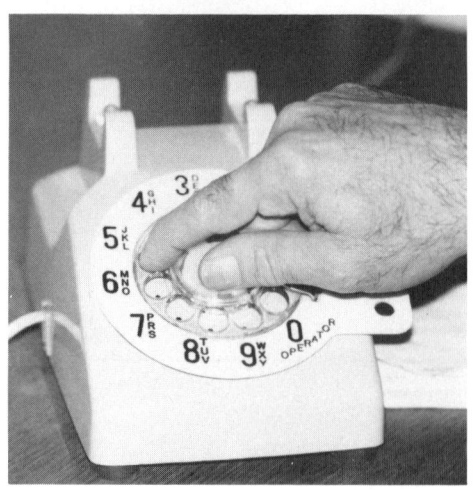

A number of simple devices have been designed to aid the visually impaired. Several are shown here. *Top left, clockwise:* A line guide for filling out checks, a cooking timer and large raised dots on a domino game, and a guide with enlarged numbers for dial phones.

Developing products for the handicapped goes beyond design considerations, however, according to OTA. In a 1982 study "Technology and Handicapped People," OTA found that the tools are available to develop the technological means to help the disabled function effectively in the mainstream of American life. But programs for the handicapped are not coordinated and have conflicting and ill-defined goals. In addition, financing is a major barrier.

OTA found that disabled people depend on public and nonpublic programs to provide various technological aids. This is partly because many disabled people have lower-than-average earnings and partly because the programs have become the main sources for the technology and products that help the handicapped. There appears to be no mass market for manufactured goods designed for the disabled, so businesses have little interest in large-scale production. Thus, costs of technological aids remain high and often out of reach for many disabled.

Even when handicapped people can afford to buy equipment they need, they may still have problems obtaining new or experimental products. That was the case for a San Francisco woman, Carol Raugust, who is confined to a wheelchair. She tried to bring into the United States a box-like British car, the Elswick Envoy. The experimental car, specifically designed for the disabled, would allow Ms. Raugust to stay in her wheelchair while driving, and to get in and out without help. But California blocked entry of the vehicle because it does not meet emission control standards set by the state's Air Resources Board.

Although Ms. Raugust, a microbiologist, well appreciates the need to guard against air pollutants, she also thought people like herself should be allowed to use a vehicle that makes them more "able." With the help of a state legislator and a church group which raised funds for the car, Ms. Raugust campaigned for an exemption to the law, which was at first denied. But her case was widely publicized, and in March 1984 the Air Resources

Board allowed Ms. Raugust to import the car "for experimental reasons" only. The exemption does not apply to any other persons or vehicles.

Very few American cars have been designed specifically for drivers in wheelchairs, but many mass-produced cars have been modified for various types of handicaps. And a variety of vehicles such as motorized carts and tricycles have been developed for the disabled. A Chair-E-Yacht is also designed for a wheelchair-bound person. This small open van with a ramp holds only a wheelchair and its occupant, but is designed for longer traveling distances than a battery-powered wheelchair will allow.

One of the most revolutionary developments for the handicapped has been the use of the computer. Many computer programs and computerized devices and robots have been and are being developed to make life more "ordinary" for people with functional limitations. Some computer programs, for example, teach living skills. Such a program used in a New York school for the deaf shows a replica of a local supermarket. Students use it to learn how to locate various products on the shelves, how to check prices, and how to add up costs. After working with the simulation, students are able to shop with little difficulty in the real supermarket. Similar, but more simplified, programs are used in classes for the trainable mentally retarded.

Computers are also helping people who are unable to speak, or have limited movement. Adaptive devices such as mouth sticks and head wands help people tap out messages on computer keyboards. Some computer programs can be operated by voice commands. Others change a printed page into synthesized speech.

Artificial limbs—arms and legs—have been developed to operate with the body's own electrical signals that are picked up by microcomputers. The microcomputers convert the signals to commands that make battery-powered limbs move—opening and closing a hand, for example.

HANDIABLE DESIGNS / 103

Research is underway to develop a walking machine for people who are paralyzed because of damage to the spinal cord. This device also depends on a microcomputer that picks up electrical signals from muscles. Then the computer sends commands to an electrostimulator. The stimulator in turn activates leg muscles that are paralyzed.

Many computer devices have been designed to fit the needs of the blind or near blind. Some print in Braille or in raised letters. For the visually impaired, products range from a "talking clock," which gives the time in a distinct voice and can be programmed to tell the time every hour, to a "say when" indicator, a small electronic device that fits over the edge of a cup or glass and buzzes when the container is filled to a half inch from the top.

People with very limited vision are unable to clearly see the characters on most computer screens, which can be a serious problem for the partially sighted students who are "mainstreamed" in classrooms. But the large print display processors from Visualtek in Santa Monica, California, help the partially sighted use computers as efficiently as the fully sighted.

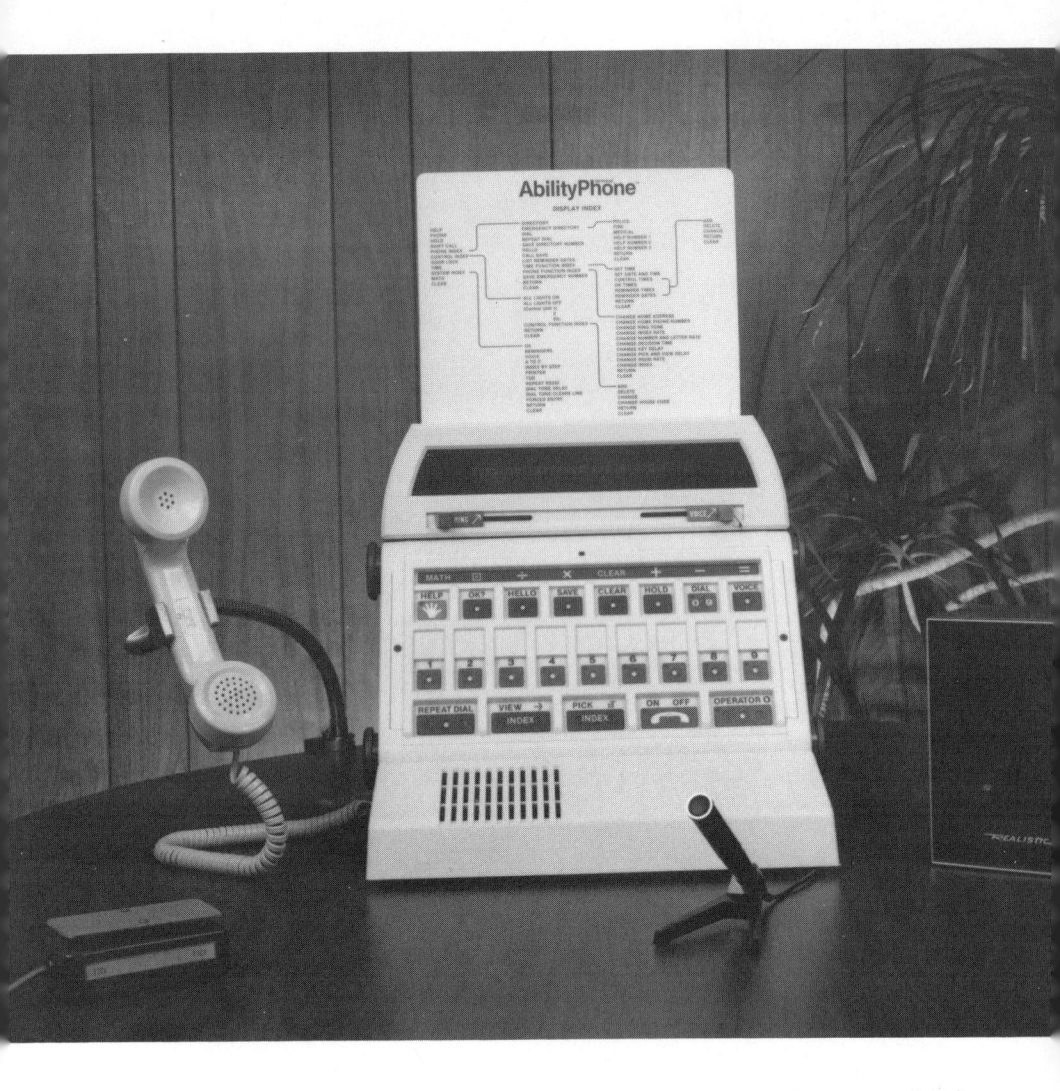

The AbilityPhone, designed by Tom Cannon, a human factors specialist in Fort Collins, Colorado, was "developed after an extensive study of the human factors problems encountered by disabled persons in living and working independently," Cannon says. The computerized terminal was developed to meet the needs of 100 percent of the population, which means that the severely disabled as well as the ablebodied population can use the device.

The terminal assists people who happen to be handicapped achieve an independent life. For example, severely disabled individuals can make and receive phone calls, using special switches, if necessary, that can be operated by various parts of the body or by breath control. A user can initiate an automatic call for help or safety devices such as a smoke detector will sound an alarm. The terminal has also been designed to control lights, turn appliances on or off, or to make calculations through voice or large keyboard controls.

A number of products designed for the general population have been adapted or modified for use by people with functional limitations. In recent years, buildings, public restrooms, walkways, and transportation systems have been modified so that disabled people can use these facilities. Modifications are especially helpful in the workplace, whether an industrial site or office.

John Mueller, a human factors consultant in New York, developed a workbook to help employers modify worksites to fit various functional limitations. The designs, created while Mueller was with the George Washington University Rehabilitation Research and Training Center in Washington, D.C., show two typical worksites: an office and industrial setting. Each site diagram (page 106) includes design modifications that would allow many disabled people to function effectively.

Mueller also designed modifications to fit the needs of people with specific disabilities. For example, persons who have difficulty with arm movements—about 6 percent of the U.S. population—may need light switch extensions, mechanical "reachers," self-correcting typewriters, telephone aids such as a headset receiver, and desk-top organizers. People with sight limitations may need lighted magnifiers, textured controls of various kinds, strips on the floor to indicate an exit, lever door handles, "talking" calculators, and video reading aids.

When designing or modifying worksites to fit people with functional limitations, an employer should consult with each disabled person, Mueller says. Then the worksite will fit the individual. The result, he believes, "will be a more productive environment for disabled and nondisabled as well."

Although the elderly are not necessarily disabled, many older people can benefit from built space and products designed for their needs and safety. Research shows that people over age fifty-five have many more product-related accidents than the

OFFICE WORKSITE

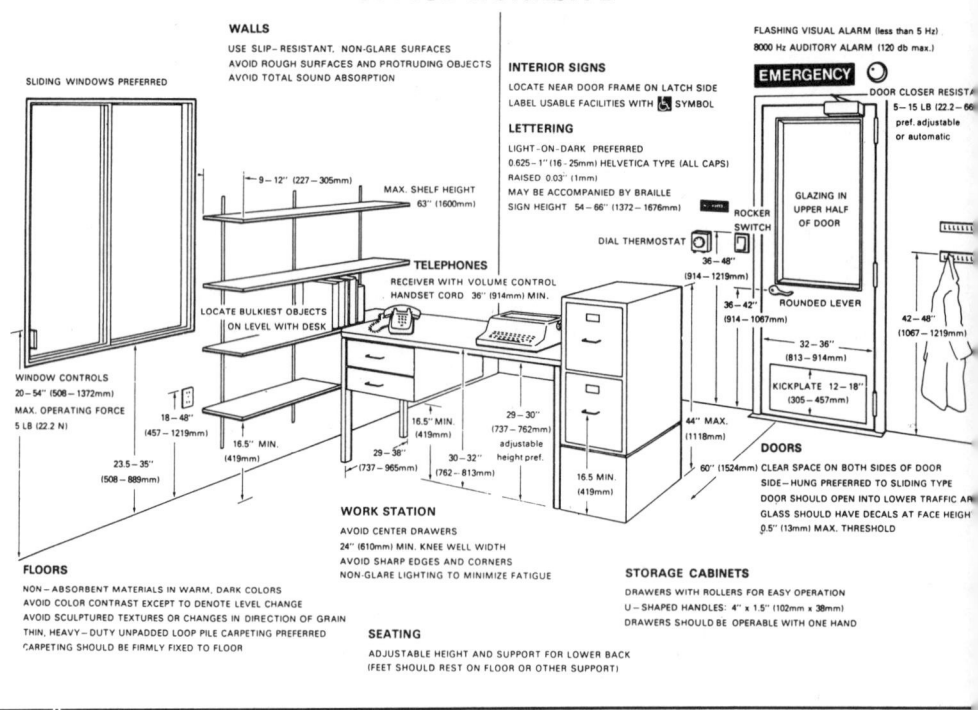

WALLS
USE SLIP-RESISTANT, NON-GLARE SURFACES
AVOID ROUGH SURFACES AND PROTRUDING OBJECTS
AVOID TOTAL SOUND ABSORPTION

INTERIOR SIGNS
LOCATE NEAR DOOR FRAME ON LATCH SIDE
LABEL USABLE FACILITIES WITH ♿ SYMBOL

LETTERING
LIGHT-ON-DARK PREFERRED
0.625–1" (16–25mm) HELVETICA TYPE (ALL CAPS)
RAISED 0.03" (1mm)
MAY BE ACCOMPANIED BY BRAILLE
SIGN HEIGHT 54–66" (1372–1676mm)

FLASHING VISUAL ALARM (less than 5 Hz)
8000 Hz AUDITORY ALARM (120 db max.)

EMERGENCY

DOOR CLOSER RESISTA
5–15 LB (22.2–66
pref. adjustable
or automatic

SLIDING WINDOWS PREFERRED

9–12" (227–305mm)
MAX. SHELF HEIGHT 63" (1600mm)

TELEPHONES
RECEIVER WITH VOLUME CONTROL
HANDSET CORD 36" (914mm) MIN.

DIAL THERMOSTAT

ROCKER SWITCH

GLAZING IN UPPER HALF OF DOOR

36–48" (914–1219mm)

LOCATE BULKIEST OBJECTS ON LEVEL WITH DESK

36–42" (914–1067mm) ROUNDED LEVER

32–36" (813–914mm)

42–48" (1067–1219mm)

KICKPLATE 12–18" (305–457mm)

WINDOW CONTROLS
20–54" (508–1372mm)
MAX. OPERATING FORCE
5 LB (22.2 N)

18–48" (457–1219mm)

16.5" MIN. (419mm)

29–30" (737–762mm) adjustable height pref.

44" MAX. (1118mm)

23.5–35" (508–889mm)

16.5" MIN. (419mm)

29–38" (737–965mm)

30–32" (762–813mm)

16.5 MIN. (419mm)

DOORS
60" (1524mm) CLEAR SPACE ON BOTH SIDES OF DOOR
SIDE-HUNG PREFERRED TO SLIDING TYPE
DOOR SHOULD OPEN INTO LOWER TRAFFIC AR
GLASS SHOULD HAVE DECALS AT FACE HEIGH
0.5" (13mm) MAX. THRESHOLD

WORK STATION
AVOID CENTER DRAWERS
24" (610mm) MIN. KNEE WELL WIDTH
AVOID SHARP EDGES AND CORNERS
NON-GLARE LIGHTING TO MINIMIZE FATIGUE

STORAGE CABINETS
DRAWERS WITH ROLLERS FOR EASY OPERATION
U-SHAPED HANDLES: 4" x 1.5" (102mm x 38mm)
DRAWERS SHOULD BE OPERABLE WITH ONE HAND

FLOORS
NON-ABSORBENT MATERIALS IN WARM, DARK COLORS
AVOID COLOR CONTRAST EXCEPT TO DENOTE LEVEL CHANGE
AVOID SCULPTURED TEXTURES OR CHANGES IN DIRECTION OF GRAIN
THIN, HEAVY-DUTY UNPADDED LOOP PILE CARPETING PREFERRED
CARPETING SHOULD BE FIRMLY FIXED TO FLOOR

SEATING
ADJUSTABLE HEIGHT AND SUPPORT FOR LOWER BACK
(FEET SHOULD REST ON FLOOR OR OTHER SUPPORT)

INDUSTRIAL WORKSITE

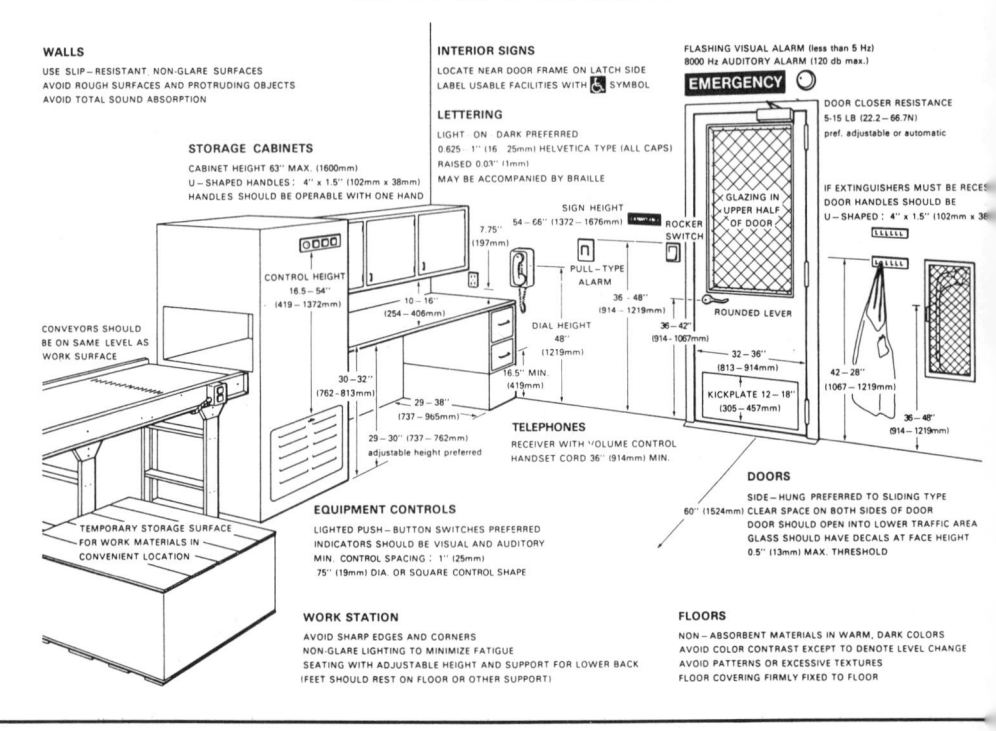

WALLS
USE SLIP-RESISTANT, NON-GLARE SURFACES
AVOID ROUGH SURFACES AND PROTRUDING OBJECTS
AVOID TOTAL SOUND ABSORPTION

INTERIOR SIGNS
LOCATE NEAR DOOR FRAME ON LATCH SIDE
LABEL USABLE FACILITIES WITH ♿ SYMBOL

LETTERING
LIGHT-ON-DARK PREFERRED
0.625–1" (16–25mm) HELVETICA TYPE (ALL CAPS)
RAISED 0.03" (1mm)
MAY BE ACCOMPANIED BY BRAILLE

SIGN HEIGHT 54–66" (1372–1676mm)

FLASHING VISUAL ALARM (less than 5 Hz)
8000 Hz AUDITORY ALARM (120 db max.)

EMERGENCY

DOOR CLOSER RESISTANCE
5–15 LB (22.2–66.7N)
pref. adjustable or automatic

STORAGE CABINETS
CABINET HEIGHT 63" MAX. (1600mm)
U-SHAPED HANDLES: 4" x 1.5" (102mm x 38mm)
HANDLES SHOULD BE OPERABLE WITH ONE HAND

7.75" (197mm)

ROCKER SWITCH

GLAZING IN UPPER HALF OF DOOR

IF EXTINGUISHERS MUST BE RECES
DOOR HANDLES SHOULD BE
U-SHAPED: 4" x 1.5" (102mm x 38

CONTROL HEIGHT
16.5–54" (419–1372mm)

10–16" (254–406mm)

PULL-TYPE ALARM

36–48" (914–1219mm) ROUNDED LEVER

CONVEYORS SHOULD BE ON SAME LEVEL AS WORK SURFACE

30–32" (762–813mm)

29–38" (737–965mm)

DIAL HEIGHT 48" (1219mm)

36–42" (914–1067mm)

32–36" (813–914mm)

42–28" (1067–1219mm)

16.5" MIN. (419mm)

KICKPLATE 12–18" (305–457mm)

36–48" (914–1219mm)

29–30" (737–762mm) adjustable height preferred

TELEPHONES
RECEIVER WITH VOLUME CONTROL
HANDSET CORD 36" (914mm) MIN.

TEMPORARY STORAGE SURFACE FOR WORK MATERIALS IN CONVENIENT LOCATION

DOORS
SIDE-HUNG PREFERRED TO SLIDING TYPE
60" (1524mm) CLEAR SPACE ON BOTH SIDES OF DOOR
DOOR SHOULD OPEN INTO LOWER TRAFFIC AREA
GLASS SHOULD HAVE DECALS AT FACE HEIGHT
0.5" (13mm) MAX. THRESHOLD

EQUIPMENT CONTROLS
LIGHTED PUSH-BUTTON SWITCHES PREFERRED
INDICATORS SHOULD BE VISUAL AND AUDITORY
MIN. CONTROL SPACING : 1" (25mm)
.75" (19mm) DIA. OR SQUARE CONTROL SHAPE

WORK STATION
AVOID SHARP EDGES AND CORNERS
NON-GLARE LIGHTING TO MINIMIZE FATIGUE
SEATING WITH ADJUSTABLE HEIGHT AND SUPPORT FOR LOWER BACK
(FEET SHOULD REST ON FLOOR OR OTHER SUPPORT)

FLOORS
NON-ABSORBENT MATERIALS IN WARM, DARK COLORS
AVOID COLOR CONTRAST EXCEPT TO DENOTE LEVEL CHANGE
AVOID PATTERNS OR EXCESSIVE TEXTURES
FLOOR COVERING FIRMLY FIXED TO FLOOR

rest of the population. The death rate from accidents is twice that of other age groups.

One of the most hazardous products is flooring or flooring materials (such as a slick linoleum). Lawn mowers, ladders, bathtubs, and showers are other products that are common hazards for older people. Many items in these categories could be designed or modified to include more safety features.

When designing built space for the elderly, such as retirement homes, builders need to consider factors that allow occupants to function independently in living quarters. For example, windows should open and close easily. Furnace, stove, and refrigerator controls should be highly visible and easy to understand. The bathtub or shower should have support rails to help prevent falls. The building itself, as well as individual living quarters, should be barrier-free for those who must use wheelchairs or other aids for mobility. Alarm devices, both visual and auditory, should be designed so that people with sight or hearing limitations will be adequately warned in an emergency.

Another area that concerns human factors researchers is the difficulty some older people and disabled persons have opening various types of containers without help. The problem is evidently widespread, since dozens of devices have been manufactured to help people open or twist off lids. A recent human factors study, sponsored by the University of Kansas Gerontology Center in Kansas City, Kansas, found that much could be done to improve the design of jars and lids on various products. Many older people have difficulty opening jars because they do not have enough strength in the wrist to twist off lids.

In one of the tests, human factors researchers measured the wrist-twisting strength of 100 men and 100 women between the ages of sixty-two and ninety-two. Using a modified torque-wrench device developed by the researchers, the volunteer subjects tested their abilities to twist eight different container lids. The types of food containers used included jars of peanut butter, instant coffee, olives, and salad dressing, and a soft drink bottle.

Researchers found, as they expected, that elderly women had more difficulty than men in twisting off lids. But, overall, the tests showed that "short jars with large lids would probably be more widely accepted [by the elderly] than the current and almost universal tall jar with a small diameter."

As the population in the United States grows older—by the year 2033 more than one of every five people will be elderly—more attention will be paid to human factors associated with older consumers and users of built environments. Research, for example, is already being conducted on the effects of age on attention span as needed for certain job tasks. Other studies are attempting to determine what types of housing, medical facilities, and leisure products will be needed by the elderly. Whatever the findings, the results should benefit a wide variety of individuals—those with limitations imposed by aging or by functional disabilities, and those with so-called "normal" abilities.

TOOLS ERGONOMISTS USE 10

This is just a partial list of the many instruments and devices ergonomists use to determine physiological responses to various tasks and how work areas and products fit people.

An anthropometer to measure various body dimensions.

A sliding caliper to measure the distance across the hand.

A goniometer to measure the wrist angle and range of motion of the wrist.

An ergometer to measure physical work.

An illumination meter to measure the amount of light falling on a surface.

A sound-level meter to measure sound pressure levels.

A three-dimensional grid to measure arm reach.

A biotelemetry transmitter to measure heart rate.

An oxygen analyzer to measure the consumption of oxygen.

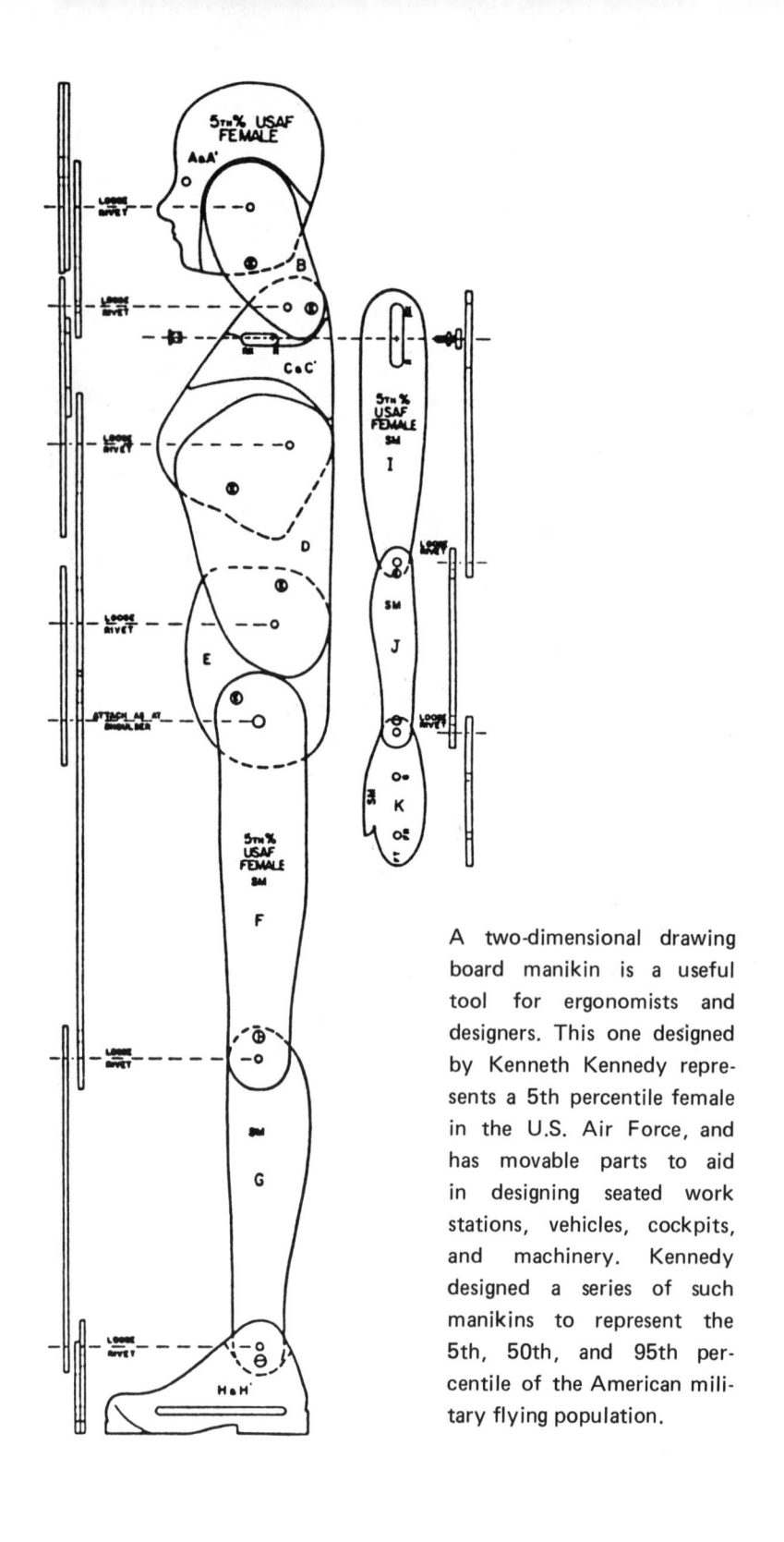

A two-dimensional drawing board manikin is a useful tool for ergonomists and designers. This one designed by Kenneth Kennedy represents a 5th percentile female in the U.S. Air Force, and has movable parts to aid in designing seated work stations, vehicles, cockpits, and machinery. Kennedy designed a series of such manikins to represent the 5th, 50th, and 95th percentile of the American military flying population.

This three-dimensional grid represents the reach capability of a seated operator with short arms (5th percentile). With such a model, a workstation can be designed so that an operator does not have to lean forward or stretch continually to accomplish a task, thus reducing fatigue. (From the book *Ergonomic Design for People at Work,* Vol. 1, by Eastman Kodak Company. Copyright Eastman Kodak Company, 1983. Used by permission.)

A number of other tools and techniques also help ergonomists gather and analyze data. An important one is the questionnaire. For example, to evaluate a product, human factors specialists have to design questions that are easily understood by those filling out the questionnaire. At the same time, the questions must bring responses that tell manufacturers what they need to know to analyze a product, such as how it is used, whether there are any health or safety hazards, and what features might need improvement.

A procedure widely used by ergonomists in industry is the workplace survey or workplace evaluation. Several human factors specialists or a team of ergonomists and health and safety personnel in a manufacturing plant conduct surveys of work areas or job sites. Supervisors usually provide the initial information on possible human factor, safety, or health problems. A supervisor may have observed many worker errors, that absenteeism is high, that production may have dropped, or that there have been a number of injuries on the job. The supervisory people may suggest when and where a survey should be made and may prepare workers for interview questions, explaining why an evaluation is needed and what improvements might be made as a result of the information gathered.

Checklists help pinpoint problems that might not be obvious to workers, designers, or even to the ergonomists who are conducting workplace studies. Usually the checklist covers a variety of questions about the design of the workspace, such as whether the seating and lighting are adequate for the tasks to be done. Other checkpoints have to do with the use of tools and machinery, the ease of reading displays and operating controls, environmental conditions, organization, and methods of work.

Ergonomists also use other types of evaluation in their attempts to match products to people. As described in previous chapters, volunteers test products, sometimes experimenting on their own and reporting their reactions. But more often, human factors specialists set up experiments so that volunteers can be interviewed and filmed on video tape while using products.

A simulation is one more method ergonomists use to gather information. If a workstation or task is under study, a replica of the work area might be set up to analyze, with volunteer help, how the station fits a worker or how the task is performed. At several Kodak plants, for example, proper lighting is important for more than a hundred different jobs. At one type of station, inspectors have to carefully examine photographic paper made

Nick Simonelli (right), a human factors specialist with 3M Company, interviews a volunteer for her reactions on how easy or difficult a tape product is to use. The interview is being filmed on video tape for later analysis.

by the firm. To control the quality of the product, inspectors must detect and report on any minute flaws in the paper. The job requires more than good eyesight and plenty of bright light. The angle of the light, the distance of the light from the product, and the effects of shadows and color are also important considerations.

In the Human Factors Laboratory at Kodak, experiments were conducted with a model, or mockup, of the paper inspection station. Using a simple wooden frame, various types of lights of different intensities were placed at different angles to the model inspection table. With the aid of volunteers, the human factors staff tested and analyzed the effects of the different lighting arrangements and were able to determine the best kind of light for inspection jobs and other work areas as well.

Human factors specialists at Kodak often design monitoring devices such as this one worn by the woman at the right whose wrist motions are being recorded by the designer of the experiment, Dick Little (left). (From the book *Ergonomic Design for People at Work,* Vol. 1, by Eastman Kodak Company. Copyright Eastman Kodak Company, 1983. Used by permission.)

Computer graphics is an important tool for human factors engineers. Developed from space technology, the images are produced by computer-processed signals and are displayed on a screen in picture form. Using computer commands, engineers and designers can create graphics to illustrate, for example, the exact dimensions of a new instrument or machine, or they can see in graphic, animated form whether a workstation or job task will be suitable for users. Perhaps a job causes high physical stress. A computer can simulate with graphics the stresses on the body and thus aid in the design of work tasks or workstations that reduce stress.

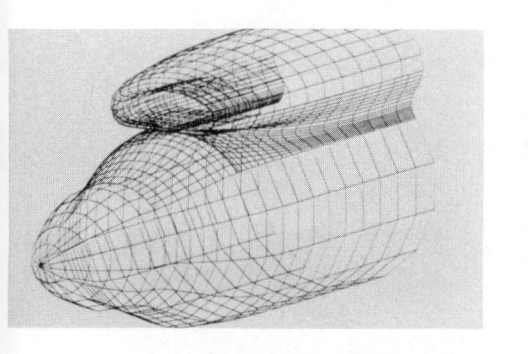

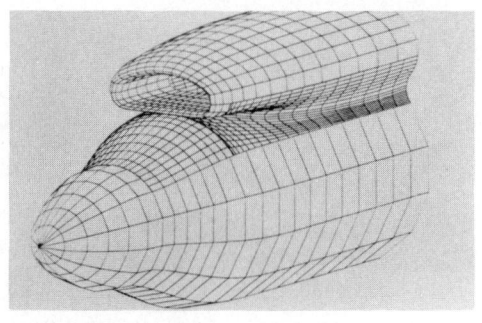

Computer-generated drawings, such as these shown above, are important tools for ergonomists. The first drawing, left, was generated by a computer at Lockheed-Georgia Company, which was developing an advanced turboprop aircraft. However, the drawing shows all the lines that result because of data input. A new computer program developed by NASA now produces a cleaner drawing, right, that shows perspective.

Computer simulation combined with mockups of workstations are also valuable ergonomic tools, especially when complex designs are under study. At King's Point (New York) Merchant Marine Academy, a simulator for use in human factors research of cargo ship handling and safety is a realistic mockup of a 14- by 20-foot ship's bridge. It includes equipment and instruments found on cargo ships, complete with windows surrounding the pilot work area. Through the windows, a ship's pilot on the bridge can see scenes of a waterway and shoreline—full-color pictures generated by a computer and projected onto a screen that provides a 240-degree view.

The views of buildings, piers and other structures change as the pilot manipulates the controls and simulates guiding a huge 80,000- to over 200,000-ton tanker through a harbor. There is a sense of motion because the computer also causes the mockup to simulate changes in the ship's speed, the effects of waves and currents, and even of different weather conditions.

During such simulations, ergonomists can observe via TV monitors how test subjects (volunteer ship masters) react. Researchers use data from the tests to make recommendations as to cargo ship bridge designs, how to reduce fatigue for crew members standing watch, and what measures can be taken to reduce the possibility of collisions or groundings of huge ships.

Mockups and computer simulations are common in aerospace research and design. A mockup of a space station control center with computer simulations has recently been completed at the Johnson Space Center. The mockup simulates the interior of a space station command center including computer keyboards and display screens for monitoring space station systems. NASA engineers are using the mockup to develop and test the computer technology that will control stations in space.

According to human factors specialists at the U.S. Army Research Institute, Ft. Leavenworth, Kansas, "The U.S. Army is rapidly expanding its use of computer simulations to provide [battlefield] command and control training." And computer simulations have long been part of various types of military and commercial flight training.

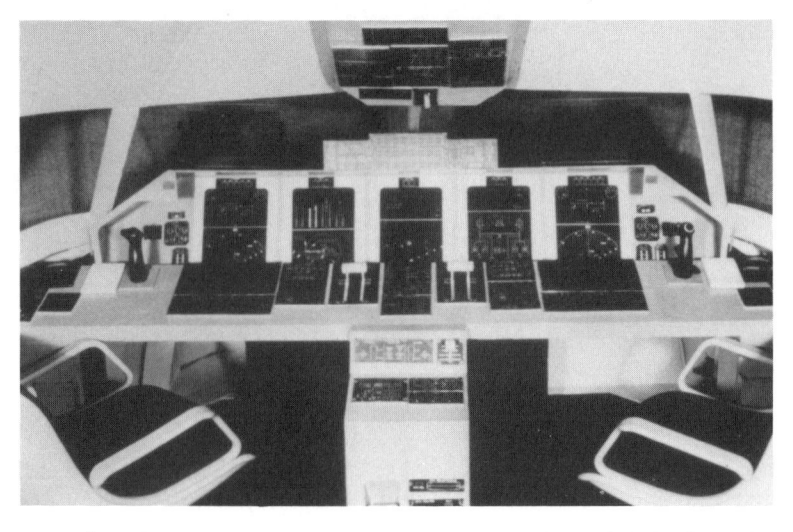

A simulated flight deck of a commercial transport that might be in service in the 1990s. It can run realistic full-mission, full-crew simulations of transport operations. The results of such tests will provide guidelines for further development and certification of advanced automated cockpit systems.

This hypobaric altitude chamber manufactured by Nautilus Environmedical Systems is used for flight training by the military. Inside the chamber, normal atmospheric pressure is reduced to simulate airborne flight operation. Chamber occupants breathe oxygen and communicate via masks, helmets, and console systems. A flight director can monitor trainees through the windows. The physiological reactions of occupants to the reduction of pressure and the psychological reactions to explosive decompression are measured and evaluated.

Whether for use in simulations, storing and analyzing data, or as educational aids, computers are valuable tools for human factors researchers. At the same time that ergonomists work on the designs of computer systems and computer products to make them more "user-friendly," they also make use of computer databases to locate studies, statistics, and other information available on human factors design problems and how human factors research can be applied.

The explosion of electronic information technology is both a benefit and a source of frustration to ergonomists. As Frederick Muckler, chief scientist/vice president of Essex Corporation, put it, "Every year, hundreds (if not thousands) of research reports relevant to human factors are generated throughout the world. It is not so much a question of the time to read them as it is to know that they exist." The primary difficulties, as Muckler sees them, are distributing information so it can be used by human factors specialists and to communicate human factors concerns to the general public.

Public awareness of ergonomic issues may prove to be one of the most important "tools" in terms of expanding the work of ergonomists. As people learn more about user-oriented products and workstations, they will demand that more attention be paid to designs with people in mind. In turn, this could mean much broader application for human factors research—in many areas including agriculture, law enforcement, and health facilities. Dr. Muckler summed it up for the Human Factors Society when he noted, "The potential for new applications of human factors is unlimited. Wherever there are humans performing tasks by whatever tools, there is a role for human factors in helping to design and perform those tasks."

GLOSSARY

anthropometric—having to do with measurements of the human body.

automation—use of electronic or mechanical devices to control or operate machines.

"built space"—indoor or outdoor areas built for a specific purpose, such as a playground.

electronics—the use of the energy of electrons in machines and devices.

ergonomics—from the Greek *ergon* (work) and *nomos* (laws), thus "the laws of work"; an applied science that deals in part with making workplaces fit people.

gerontology—the study of aging in people.

"handiable"—being able to function independently or with limited aid even though disabled.

neurological—having to do with the nervous system or the study of the nervous system.

novice—a beginner.

ophthalmologist—a physician who deals with the functions and diseases of the eye.

physiological—having to do with the functions of living things.

120 / GLOSSARY

synthesized—parts or elements combined to make a whole; an artificial treatment.

technology—the application of scientific research, particularly in industry.

toxicity—poisonous quality.

VDT—video display terminal: a computer screen.

vertebrae—bones making up the spinal column.

ABBREVIATIONS

DOT	Department of Transportation
EPA	Environmental Protection Agency
NASA	National Aeronautics and Space Administration
NIOSH	National Institute for Occupational Safety and Health
NRC	Nuclear Regulatory Commission
OSHA	Occupational Safety and Health Administration
OTA	Office of Technology Assessment
UAW	United Automobile, Aerospace, and Agricultural Implement Workers of America

ADDRESSES

Aerospace and Human Factors Division
Ames Research Center
National Aeronautics and Space
 Administration (NASA)
Moffett Field, CA 94035

American Industrial Hygiene Association (AIHA)
Public Relations Office
475 Wolfledges Parkway
Akron, OH 44311

Center for Ergonomics
University of Michigan
1205 Beal
Ann Arbor, MI 48109

Consumer Product Safety Commission
Human Factors Division
5401 Westbard Avenue
Bethesda, MD 20207

Human Factors Safety Division
Nuclear Regulatory Commission (NRC)
Washington, D.C. 20555

Human Factors Society, Inc. (HFS)
P.O. Box 1369
Santa Monica, CA 90406

Institute of Industrial Engineers
25 Technology Park Atlanta
Norcross, GA 30092

National Transportation Safety Board
Office of Government and Public Affairs
800 Independence Avenue, SW
Washington, D.C. 20594

Special Interest Group on Computers
 and Human Interaction (SIGCHI)
Association for Computing Machinery
11 West 42nd Street
New York, NY 10036

Technical Information Branch
National Institute for Occupational
 Safety & Health (NIOSH)
4676 Columbia Parkway
Cincinnati, OH 45226

FURTHER READING

BOOKS

Black, Stephen, *Man and Motor Cars; An Ergonomic Study*. New York: Norton, 1967.

Edholm, Otto Gustaf, *The Biology of Work*. New York: McGraw-Hill, 1968.

Engle, Eloise and Lott, Arnold, *Man in Flight: Biomedical Achievements in Aerospace*. Annapolis, Md.: Leeward Publications, Inc., 1979.

Ergonomics Guidebook. National Safety Council, 444 North Michigan Avenue, Chicago, Ill. 60611.

Ergonomics Handbook. International Business Machines Corporation, Armonk, N.Y. 10504.

Hammond, John, *Understanding Human Engineering: An Introduction to Ergonomics*. North Promfret, Vt.: David & Charles, Inc., 1979.

Lucie-Smith, Edward, *A History of Industrial Design*. New York: Van Nostrand Reinhold Co., 1983.

McCormick, E.J. and Sanders, M.S., *Human Factors in Engineering and Design*. 5th ed. New York: McGraw-Hill, 1982.

Osborne, David J., *Ergonomics at Work*. New York: John Wiley & Sons, Inc., 1982.

Sommer, Robert, *Social Design: Creating Buildings with People in Mind*. Englewood Cliffs, N.J.: Prentice-Hall, Inc., 1983.

Strains & Sprains: A Worker's Guide to Job Design. International Union, United Automobile, Aerospace, and Agricultural Implement Workers of America, UAW, Health and Safety Department, 800 E. Jefferson, Detroit, Mich. 48214.

Technology Utilization and Industry Affairs Division, NASA, *Spinoff.* Washington, D.C.: U.S. Government Printing Office (annual).

Van Cott, H.P. and Kinkade, R.G., Eds., *Human Engineering Guide to Equipment Design* (revised edition). Washington, D.C.: U.S. Government Printing Office, 1972.

PERIODICALS

Barry, V.T.R., "Kidspace: Family Life in the City," *Children Today,* July-August, 1982, pp. 11-15.

Begley, Sharon, et al, "Making Machines Fit People," *Newsweek,* August 29, 1983, p. 68.

Campbell, Bruce, "Comfort by Design," *Working Woman,* May 1984, pp. 102-104.

Cirillo, David J., "Office Ergonomics: Coping with Causes of Stress in the Automated Workplace," *Management Review,* December 1983, pp. 25- .

Dixon, Bernard, "That Sinking Feeling: Ergonomic Design Fails to Impress Bernard Dixon," *New Scientist,* August 1984, pp. 41-42.

"Enter Ergonomics," *Current Health,* November 1984, pp. 12-13.

Goldfield, Randy, "New Office Ergonomics," *Working Woman,* August 1983, pp. 50-52.

Kaercher, Dan, "Amazing Advances in Medical Technology," *Better Homes and Gardens,* January 1984, pp. 15-19.

Karmin, Monroe W., et al, "High Tech: Blessing or Curse?" *U.S. News and World Report,* January 16, 1984, pp. 38-44.

McQuade, Walter, "Easing Tensions Between Man and Machine," *Fortune*, March 19, 1984, pp. 58-66.

Micossi, Anita, "Ergonomics Is Good Business," *PC World*, December 1983, pp. 300-307.

Rice, Berkeley, "Don't Touch That Dial," *Psychology Today*, January 1984, pp. 64-67.

Rosenfeld, Neill, "Must Machines Be Anti-Human?" *Newsday*, November 1, 1981, pp. 12-

Self, Jim, "Ergonomics and Productivity in the Electronic Office," *Legal Administrator*, Fall 1983, pp. 18-23.

Shell, Ellen Ruppel, "Designing for the Errant Elbow," *Technology Illustrated*, May 1983, pp. 54-58.

Sheridan, Thomas B., "Human Error in Nuclear Power Plants," *Technology Review*, February 1980, pp. 22-23.

Weissler, Paul, "Designing Cars to Fit People," *Mechanix Illustrated*, July 1982, pp. 46-

INDEX

A
air bags, 70-73
American Newspaper Guild, 31
American Telegraph and Telephone, 15, 54
anthropometric data, 34, 36, 69, 85, 86
Anthropometric Source Book, 85
Apollo missions, 78
Armstrong, David, 33
artificial limbs, 102
Association of Ophthalmologists, 32

B
balloonist, 77
Bell/Bell systems, 15, 54-57
"bigfoot," 69
biomechanical studies, 22
Black, Stephen, *Man and Motor Cars*, 70
Blue Shield (insurance), 31
Boeing Aerospace and Airline Companies, 15
Braille, 103

"built space," 92, 93, 98, 105, 107
Bureau of Radiological Health, 32

C
California
  Air Resources Board, 101
  Santa Barbara, 26
  Santa Monica, 12
Cannon, Tom, 99, 100
carpal tunnel syndrome, 20
Casey, Stephen, 26, 28
Center for Consumer Research, 93
Chair-E-Yacht, 102
Chapanis, Alphonse, 49
Charles Mauro Associates, 25
Chicago O'Hare International Airport, 25
color-coding, 46
computer(s), 11, 14, 31. *See also* video display terminal
  databases, 118
  handicapped and, 102, 103
  operators, 32, 37
  printers, noise from, 96
  simulation, 66, 114, 115, 116

systems, 30, 118
technology, 30, 37
"user-friendly," 31, 37, 118
workstations, 32, 33, 34
Consumer Product Safety Act, 62
Control Data Corporation, 16
controls, 24-30, 48, 49, 69, 70
Convair, 25, 26

D
Damask, Arthur, 65
Davis, Harry L., 14, 15
Davis, Sir Robert H., 77
Department of Health and Human Services, 33
Department of Transportation (DOT), 67, 72, 73
Dreyfus, Henry, 52, 55
Duke University, 32

E
Eastman, Kodak. See Kodak
elderly, 105, 107, 108
Eli Lilly Company, 15
Elswick Envoy, 101
Engle, Eloise, *Man in Flight: Biomedical Achievements in Aerospace*, 77
Environmental Protection Agency (EPA), 96
ergonomics, definition of, 12
Essex Corporation, 118
Extravehicular Mobility Unit (EMU), 82
eyestrain, 17, 32, 33

F
Facility Management Institute, 33
Faulkner, Terrence, 58

Federal Aviation Administration (FAA), 74
Fowler, Frank, 46

G
Gatchell, Susanne, 69
Gemini flights, 78
General Electric Company, 15, 23
General Motors, 69, 70, 72
George Washington University Rehabilitation Research and Training Center, 105

H
Haldane, John Scott, 77
Harvard Medical School, 60
Hockenberry, Jack, 34
Honeywell, 16
Human Accommodation Group, 69
human factors, definition of, 12
Human Factors Laboratory, 15, 113
Human Factors Society, 12, 48, 118

I
industrial revolution, 13
Insurance Information Institute, 72
International Business Machines (IBM), 15, 57

J
Johnson and Johnson, 15, 83
Johnson Engineering Corporation, 86
Johnson Space Center, 85, 116

K

King's Point Merchant Marine Academy, 115
knobs, 20, 25-27, 48
Kodak, 14, 15, 57-59, 85, 112, 113
Kreifeldt, John, 52

L

Landisc II, 61
Laveson, Jack, 48
LeCocq, Andrew, 62
lifting, 17
lighting, 30-32, 113
Lott, Arnold. *See* Engle
Louis Harris Associates, 33
Lunar Land Rover, 86

M

Manned Maneuvering Unit (MMU), 82
Massachusetts Institute of Technology, 30
McCandless, Bruce, 82
McDonnell Douglas Astronautics Company, 83
McHugh, Michael, 65, 66
Messina, John, 64, 66
Miller, Charles, M.D., 14, 15
"moon boots," 60
"Mr. Yuk" label, 41
Mt. Sinai School of Medicine, 31
Muckler, Frederick, 118
Mueller, John, 105

N

National Academy of Sciences, 32
National Aeronautics and Space Administration (NASA), 61, 85
NASAdent, 79
safety devices, 88
space systems, 83, 116
National Institute for Occupational Safety and Health (NIOSH), 32, 36
National Research Council of Canada, 97
National Safety Council, 17
National Standards Technology, Inc., 62
National Transportation Safety Board (NTSB), 74
New York Museum of Modern Art, 59
noise (level), 26, 33, 95, 96
Nuclear Regulatory Commission, 28, 30
Nussbaum, Karen, 37

O

Occupational Safety and Health Administration (OSHA), 96
Office of Technology Assessment (OTA), 99, 101
Ortho Pharmaceutical Corporation, 83
oxygen, 13, 15

P

Paskin, Arthur, 65, 66
Pauls, Jake L., 97
Perlik, Charles A., 31
problem-solving, 14

Q

quality control, 64, 113
Queens College, 65
questionnaire, 93, 111

R

radiation, 32
radioactivity, 28
Raugust, Carol, 101
Ridge, Mark, 77
robots, 11, 23, 24, 61, 102
Rochester Institute of Technology, 61
Ryan, Joseph, 62-64

S

safety, 97. *See also* computer simulation, National Aeronautics and Space Administration, National Safety Council, Consumer Product Safety Act
  driving, 66, 70, 71, 72
  hazards, 16, 17, 22
  product design, 62, 63, 64
  public, 41, 74, 76
satellite, 60, 82, 88
Scherer Laboratories, Inc., 79
seat belts, 70-72
Sheridan, Thomas, 30
Simonelli, Nick, 52, 53
simulation(s), 66, 112, 114-118
social design, 93, 94
solar system, 83
Sommer, Robert, 93
Soviets, 78
Space Shuttle, 80-83
space station, 83, 116
space suit, 77, 78
Steelcase, Inc., 34

T

Technology Review, 30
Technology Utilization Program, 89

teflon, 60
"telepresence actuator," 61
tendonitis, 20
3M Company, 16, 52, 53
Three Mile Island (TMI), 28, 30
traffic signs, 42, 67
Trial magazine, 64
Tufts University, 52
typefaces, 44

U

UNISTIK, 86
United Auto Workers (UAW), 22
University of Kansas Gerontology Center, 107
U.S. Air Force, 25
U.S. Army, 116

V

video displays, 17, 37
video display terminal (VDT), 31-34, 36
visual pollution, 50

W

weightlessness, 83
Westinghouse Electric Corporation, 15
Working Woman, 37
workplace/workstation. *See also* computer(s)
  design of, 13, 20, 45, 49, 85, 115
  injury, 16
worksites, 105,
World War II, 13, 25, 60

X

Xerox, 15

Z

zero gravity, 77, 79

620.8　　　　　　　　　　　C. 1
Ga　　　　　　　　　　　　SCE

Gay, Kathlyn
Ergonomics: making products and places
fit people

DATE DUE

NOV 01 2011		
NOV 29 2011		

GAYLORD　　　　　　　　PRINTED IN U.S.A.